WHAT IN THE WORLD ARE

MRI SCANS?

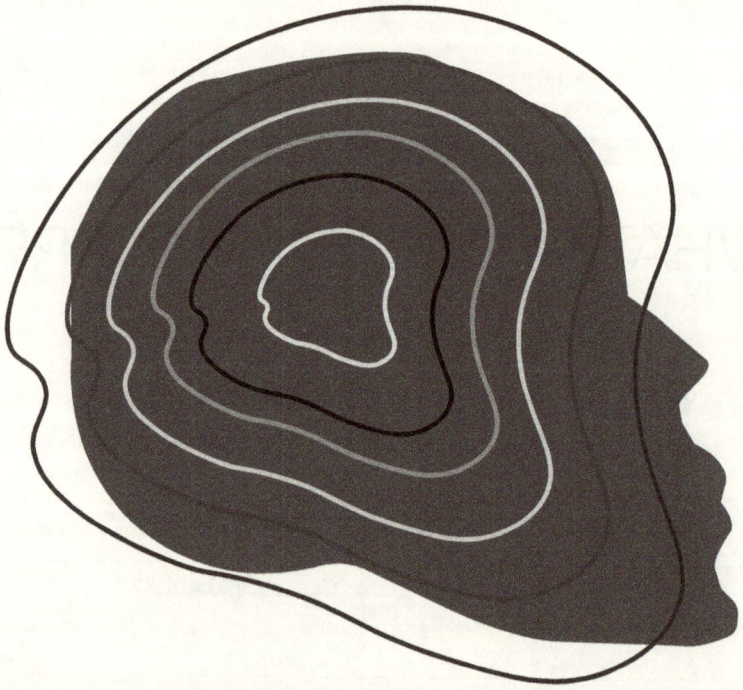

Austin Mardon   Sifar Halani

Simran Bakshi   Ashna Hudani

Shyla Bhayana   Anittha Mappanasingam

Sara Emira   Shea McMartin

Irene Fang   Rayna Kamal Naik

Jiyeon Park   Simarjit Kaur Sidhu

Mohathir Sheikh   Melissa Speagle

Edited by **Kanish Baskaran**

Typeset, Cover, and Illustration by **Ferzeen Ansari**

WHAT IN THE WORLD ARE
MRI SCANS?

GM
PRESS

Typeset, Cover, and Illustration by Ferzeen Ansari

ISBN 978-1-77369-227-2
Golden Meteorite Press
103 11919 82 St NW
Edmonton, AB T5B 2W3
www.goldenmeteoritepress.com

GM

PRESS

Table Of Contents

Introduction

The MRI scanner has been an essential diagnostic tool since its introduction in the 1970's, and has a wide variety of uses in imaging, from torn ligaments to tumors (Scatliff & Morris, 2014). This book provides a detailed look into this imaging technique, including its history, usage, evolution and relevance in society.

Introduction

CHAPTER ONE

What Was Medical Imaging Like Before MRI Scans?

Jiyeon Park

What is one of the most modern and commonly used medical imaging techniques available in the world today? It is magnetic resonance imaging, also known as MRI, which was first introduced in 1973 (Edelman, 2014). A few years later, on July 3rd, 1977, this relatively new application was fully developed and first performed on a human patient. MRI is a type of imaging technique that examines internal body structures and diagnoses possible diseases in a non-invasive way through the use of magnetic properties of atomic nuclei, as discussed in chapter 6 (Tretkoff, 2006). The purpose of this development was to provide better image resolution while medically characterizing tissues and assessing the function of systems and organs (Chang et al., 1999). MRI also helps to identify cancerous tissue to improve current medical applications (Edelman, 2014). So, what did doctors use instead of MRI before its discovery?

X-RAYS

Before MRI, X-rays were the prevailing medical imaging technology as it enabled the viewing of an individual's joints, bones and internal organs (Berger et al., 2018). They were discovered by Wilhelm Conrad Röntgen on November 8th, 1895, a scientist who conducted experiments to envisage streams of electrons by using Crookes tubes, which are glass electrical discharge tubes. Röntgen further utilized a fluorescent screen, adjusting the screen through the use of a black cardboard. During this experiment, he noticed a glimmer of visible light on the screen when the screen was placed a distance away from the tube. Röntgen then conducted another experiment, replacing the black cardboard with denser objects. After collecting results from these experiments, he started a systematic study of the new radiation called "X"-rays, which at the time was not associated with a specific meaning as he was not aware of the type of this new radiation. Later, in January of 1896, Röntgen first obtained X-ray images from his wife's hand where he was able to clearly see her wedding ring on her finger (Berger et al., 2018).

X-rays have numerous applications beyond their simple radiographic images (Berger et al., 2018). Radiography provides a two-dimensional projection of an image of specific anatomical structures through X-rays. Predominantly, these X-rays examine any changes or fractures of the skeletal muscle. The difference between the attenuation coefficient of surrounding tissue and its bones creates distinct images to classify any fractures present. The attenuation coefficient indicates how easily the X-ray beam penetrates the material. Determining the changes in bone density or consistency through radiography allows doctors to diagnose the possibility of bone cancer or osteoporosis. Another application of X-rays is angiography, which helps to visualize images of arteries to determine properties such as flow rate, lumen, size, or shape. It is often hard to differentiate the vessels from surrounding tissue, so to increase X-ray image quality, contrast agents are introduced into the bloodstream to maximize attenuation coefficients. Without contrast agents, physicians will only obtain the image of the background tissue, which does not adequately show the area of interest. Today, X-rays are one of the most well-known and widely used medical imaging techniques around the world because of their non-invasive testing methods and many clinical applications (Berger et al., 2018).

ULTRASOUND

One of the most important diagnostic procedures in health care is ultrasound imaging (National Institute of Biomedical Imaging and Bioengineering, 2016). Ultrasound is a simple and harmless medical diagnostic imaging technique that shows the distance from the probe to the boundaries (organ or tissue) along with the time measurement of how long each ultrasound echo takes to return back to the machine. Afterwards, these time and distance data help to create a two dimensional image, such as the image of a fetus (National Institute of Biomedical Imaging and Bioengineering, 2016).

Among many clinical applications of ultrasound, obstetrics and gynaecology are most common (Campbell, 2013). Ultrasound provides fetal biometry to measure fetal structures such as the size of fetal organs, ear length, foot length, head and abdomen circumference, and bone length. Based on these measurements, physicians can calculate growth predictions and fetal weight. During obstetrics appointments, couples can witness their fetal movement through ultrasound and the physician can monitor fetal activity and breathing movements. Specifically, Doppler assessment can be performed by using high frequency ultrasound waves to determine how much blood is flowing in every fetal artery, including cerebral, splanchnic and renal. The use of ultrasound further allows gynaecologists to protect women's health. Since the early stage of ovarian cancer is asymptomatic, its mortality rate is highest among all gynaecological cancers and can thus lead to devastating patient outcomes. However, with the advancement of this technology, physicians can perform screening for malignancy to determine the existence of cancerous cells in the body (Campbell, 2013). Ultrasound, thus, not only determines an individual's current health status, but also helps to prevent future possible illness.

ENDOSCOPY

In 1805, a physician named Phillip Bozzini developed a procedure to examine the orifices of the body and was able to create an instrument to see the human urethra (Spaner & Warnock, 1997). This instrument featured a light atop a tube to be inserted into body cavities. A wax candle inside of this instrument illuminated the areas to be examined and a mirror reflection aided in brightening body cavities (Figure 2). This was the first endoscope-like instrument created. In 1853, Antonin Jean Desormeaux developed a more sophisticated instrument, the endoscope, to specifically examine the bladder and the urethra. Later, Adolph Kussmaul completed a series of trials and was able to conduct esophagoscopy procedures in 1868. Although he was able to take a closer look at the live human upper digestive tract for the first time, there were some challenges such as fluid collection, reflux, and poor lighting (Spaner & Warnock, 1997). A flexible gastroscope was developed in 1932 along with a bendable tip, and experiments were continuously conducted to take photographs of the stomach of live human beings (Niwa, 2008). Tatsuro Uji finally built the gastrocamera, which enabled intragastric photography in 1950. A film cassette, light bulb and camera lens were attached to the tip of the apparatus so once the gastrocamera entered the stomach, it photographed the stomach's inner surface and allowed doctors to make a diagnosis after analyzing the pictures. After years of experiments, endoscopy has been developed to provide advanced features such as viewing the inside of the body synchronously with the videoendoscope (Niwa, 2008).

The main medical application of endoscopy is diagnosis and treatment of abdominal pathologies. It is an invasive procedure; however, patients experience less pain and operative trauma, a smaller scale of scars, and a shorter recovery and hospital stay when compared to open surgery (Haase & Maier, 2018). The advancement of endoscopy has enhanced and developed surgical procedures such as laparoscopic surgery, which is mainly used for abdominal operations (Spaner & Warnock, 1997). This surgical treatment is also a useful tool to treat rectum and colon cancer (Nandakumar & Fleshman, 2010). A laparoscopic-assisted approach is specifically advantageous for colon cancer as it maximizes a patient's health, preserving the oncological outcome for a longer period. During the early stages of laparoscopic colorectal surgery, there were some concerns regarding possible metastasis promoted by laparoscopy. However, studies have proven that patients who have undergone laparoscopic surgery have a greater chance of survival compared to those who have not (Nandakumar & Fleshman, 2010). Modern endoscopy further allows for improved characterization and detection of pathological changes within the gastrointestinal tract to either prevent or catch any malignancies in the body. Scientists can obtain microscopic analysis of surface and cellular structures *in vivo* for accurate diagnosis and therefore, physicians can provide a more personalized treatment for each patient. Further improvement and development of endoscopy will ensure more efficient diagnosis and treatment in the future (Rey et al., 2014).

COMPUTED TOMOGRAPHY

The first human computed tomography (CT) scan captured a patient's brain image in October of 1971 (Bull, 1980). The new innovator was Godfrey Newbold Hounsfield. In the late 1960s, Hounsfield was working on pattern recognition in the Electrical and Musical Industry (EMI) Limited research laboratory and an idea popped into his head. What if he used the measurements of accurate X-rays from all the various angles in a closed box? What if he collected all the information and processed it by computer? These were the beginning steps for the development of the EMI scanner. Hounsfield used a gamma ray source to obtain information via a computer and then formed a three-dimensional picture. This picture was first produced in 1968 by computer programming that calculated the values of the transmission of the rays on an object from various angles, and with accurate results. Hounsfield started to wonder how this could be used for practical applications in human and animal tissues, so he designed a machine on a lathe bed, as this bed was stable and easily available. He utilized a gamma source and the machine recorded the readings while the detector and the gamma source were moving in a linear fashion. He repeated this process from various angles by reversing the lathe. Although the quality of the picture was not great, the process still proved that the system worked. Hounsfield then collaborated with radiologists for clinical applications, leading to a great advancement in radiology. He used a brain specimen to test his principle as this was an ideal prototype to handle, and proved the validity of his principle for clinical applications. Thereafter, Hounsfield began to design a machine specifically for the brain and this prototype machine was completed in October of 1971, reducing scanning time considerably (Bull, 1980).

Today, CT provides computerized x-ray imaging by using a narrow X-ray beam. This beam is pointed at the patient's body and rotates around the body, providing cross-sectional images of the body (National Institute of Biomedical Imaging and Bioengineering, n.d.). These images are more detailed than the images obtained from conventional X-rays. The collection of this series of images, all stacked together digitally by the computer, offers a three-dimensional image of the body (National Institute of Biomedical Imaging and Bioengineering, n.d.). A CT scan, however, does not only identify large organs. It can also visualize smaller structures including calcifications and small blood vessels. High spatial resolution is used to visualize smaller structures, enabled by this advanced technology (Taubmann et al., 2018). This high-tech CT scanner thus provides high-quality data, which makes validation, quantification, and analysis possible (Zagorchev et al., 2010). Based on this information, physicians can detect anatomical abnormalities or changes in anatomy or tissue. Micro-CT, for example, offers detailed radiography and thus delivers great accuracy and reliable results, which are helpful in studying pathologic diseases, vasculature, and anatomical structures (Zagorchev et al., 2010).

One of the most valuable and widely available applications of the CT scan is in the realm of oncology because of its accurate and non-invasive procedure, suitable for the majority of diseases, and its relatively low cost (Fishman & Urban, 2001). It is critical to accurately identify whether a tumor is benign or malignant and this is possible with CT. Spiral CT is the best option for liver cancer patients as this optimal imaging procedure assists in detecting hepatic tumors. Three-dimensional images by CT allow physicians to better plan the surgery, because they can visualize real time stereoscopic images. These displays offer size measurements such as liver volume and based on the collected data, physicians can monitor the status of tumors and devise a treatment plan with more successful results. The survival rate for pancreatic cancer is very low compared to other types of cancer, so the role of CT is vital. A spiral CT scanner results in an approximately 96% tumor detection rate and indeterminate cases, including the inability to detect glandular enlargement, have substantially decreased. Since physicians can assess key venous and arterial structures, it is now possible to determine whether there is any vascular involvement within pancreatic cancer to identify tumors. The useful and accurate information gleaned from CT scans is crucial in providing hope for cancer patients (Fishman & Urban, 2001).

POSITRON EMISSION TOMOGRAPHY

Positron emission tomography, also known as PET, is a nuclear medical imaging technique that shows pathways and molecular interactions within the human body (Jones & Townsend, 2017). After the announcement of the invention of CT and Hounsfield's Nobel Prize in physiology and medicine in 1971, the value of tomographic imaging became clear and the fundamental features of CT were then used to develop PET systems (Jones & Townsend, 2017). Hoffman, Phelps, Ter-Pogossian and Mullani worked together to construct a device that could obtain reconstructed transaxial tomography, during the period 1972 to 1973. A year later, in 1974, they developed PET with an optimized scanner, whose ideal physical design eliminated any undesired coincidences (Rich, 1997). Once the PET scanner detects photons, it reassembles the image to allow visualization of spatial density to determine any change in blood flow and functional data (Portnow et al., 2013).

Although PET is an expensive procedure, it is widely used because of its beneficial applications (Das, 2015). PET is most commonly used in oncology, specifically to manage cancer patients. FDG-PET is the tracer used in PET scanning and this tracer labels lung cancer, non-Hodgkin's lymphoma, Hodgkin's disease, and other solid tumors. As a result, physicians can look for any tumor metastasis and evaluate which stage and what type of cancer is present in a patient. Personalized treatment is essential for cancer patients as each patient has a different histological diagnosis and different stage of cancer, tumour invasiveness, and sensitivity. Through PET, physicians can achieve an accurate diagnosis and therefore develop the most appropriate and personalized treatment strategy for patients, without asking for expansive biopsy surgery, in most cases. Additionally, PET scanning allows physicians to monitor the delivery of cancer therapy for the most effective outcome. Another application of PET can be found in neurological disorders. For example, dementia is usually diagnosed based on phenotypes. However, due to the overlap of phenotypes at the beginning of dementia, an inaccurate diagnosis can be made. Since F-FDG PET plays a role as a marker of neuronal injury, a PET study can reveal specific patterns of varied types of dementia (such as Alzheimer's dementia and Lewy body dementia) and therefore, physicians can differentiate the dementia type and make an accurate diagnosis (Das, 2015).

CONCLUSION

In short, the history of medical imaging has benefited from the creativity and scientific skills of many physicians and scientists. The many different forms of imaging are unique as they are all specialised and utilized for different purposes. Each new imaging technique has emerged more efficient and precise, based on previous iterations. For instance, PET provides high resolution imaging that CT was not able to generate. PET-CT is a combination of PET and CT as the PET-CT scanner has a shorter transmission time with fast CT acquisition (Das, 2015). Then, shortly after the development of PET, successful MRI was introduced to the world with a unique set of characteristics. Throughout this book, we will take a closer look at MRI, from various angles.

CHAPTER TWO

Who Invented Or Discovered MRI Scans?

Sara Emira

As with most major advancements in medical technology, the invention of MRI scans cannot be attributed to an individual researcher. Inspired by Nikola Tesla's early work on the rotation of magnetic fields, numerous scientists have dedicated their careers to adapting this concept for use in medical imaging technology. This chapter will highlight their individual contributions, the cumulative results of their findings and the scientific community's response to their achievements.

ISIDOR RABI - SETTING THE FOUNDATION

Despite the many influences on MRI technology, there is no researcher more fitting to use as a starting point than Isidor Rabi. Rabi's family immigrated from Rymanov, Austria (now located in Poland) to the US when he was only 1 year old (Shampo et al., 2012). They settled in New York City, where Isidor attended the city's public schools and thrived as a student (Shampo et al., 2012). Upon graduating, he received a scholarship to study at the prestigious Cornell University, where he majored in Chemistry (Shampo et al., 2012). Upon graduation, he spent several years working as a chemist before returning to Cornell to pursue graduate studies in Physics (Shampo et al., 2012).

After completing a Ph.D. and a postgraduate fellowship, Rabi was appointed as a professor at Columbia University where he worked for almost 40 years (Shampo et al., 2012). It was there in 1937 that he developed Nuclear Magnetic Resonance (NMR) - a method used to measure the movement of atomic nuclei (University of Washington, 2014). This technology has allowed for numerous technological advancements including - but not limited to - lasers, atomic clocks and of course, diagnostic imaging (Shampo et al., 2012). The discovery of NMR led to Rabi receiving the 1944 Nobel Prize in Physics in addition to numerous honorary doctorates, awards and memberships to prestigious scientific bodies. The specifics of Isidor Rabi's work and the theory behind Nuclear Magnetic Resonance are discussed at length in chapter 6 (*What Science is Involved in MRI Scans?*) of this book.

RAYMOND DAMADIAN, PAUL C. LAUTERBUR AND PETER MANSFIELD - PUTTING NMR IN ACTION

Isidor's discovery of NMR made way for numerous scientists to begin investigating. Amongst them was Raymond Damadian, an American physician who was inspired by his experiences as a patient after the existing imaging techniques failed to pinpoint the cause of his severe abdominal pain (Wakefield, 2000). He believed that magnetic resonance could be used to distinguish cancerous cells from their healthy counterparts, based on the former retaining more water (GE Healthcare, 2019). Theoretically, the higher concentration of hydrogen atoms present in cancerous tissue would be visible using magnetic resonance. He conducted studies on tissue samples collected from mice and in 1971, he published his paper ("Tumor Detection by Nuclear Magnetic Resonance") that supported his hypothesis in *Science* (GE Healthcare, 2019). This imaging method would later be used to create the first human scanner.

Even in its early stages, Damadian's work drew the attention of other scientists including Paul C. Lauterbur whom he met in 1971 (American Chemistry Society, 2011). At the time, Lauterbur was a graduate student investigating the chemical properties of elastics and had a budding interest in the spectroscopy technology being used by his lab at Carnegie Mellon (Lauterber, 2003). In fact, it was this technology that connected their fates, as Damadian frequented the lab at Carnegie Mellon to complete his studies on cancerous cells (American Chemistry Society, 2011). Through observing Damadian's experiments and eventually reading his paper, Lauterbur realized the potential of NMR in diagnostic imaging technology and began to actively explore this area of research (Lauterbur, 2003). More specifically, Lauterbur found himself questioning whether NMR could be used to create less invasive methods of assessing tissue *in vivo* (Lauterbur, 2003), as he had little faith in the invasive methods being transferable to humans (Wade, 2003). He hypothesized that there had to be a way of extrapolating spatial information from NMR without removing tissue from the body (Lauterbur, 2003).

With that began a shift in the scientist's career as he became the first to realize the potential of gradient magnetic fields for diagnostic scans (University of Washington, 2014). Initially, he relied on the NMR spectroscopy techniques used by chemists which utilized uniform magnetic fields to receive clear magnetic signals (Wade, 2003). Eventually, he realized that using varying magnetic fields was more effective in providing spatial information (Wade, 2003). Lauterbur began by taking images of test tubes full of water (University of Washington, 2014); one which contained heavy water and the other, regular water (Wade, 2003). The resulting image proved NMR spectroscopy to be the first method to accurately distinguish between the two subjects - a groundbreaking achievement given the human body's high water content (Wade, 2003). He then went on to use NMR to take images of peppers and clams- the first live subjects of magnetic resonance imaging (University of Washington, 2014). When asked why he specifically chose clams, Lauterbur explained that "because of the time it took to do the experiment, it had to be an animal that would lie very still" (Vogel, 2003).

Success did not come easily for Lauterbur. As much as he wanted to conduct research on the use of NMR in biomedical imaging, he struggled to secure funding due to investors not believing his theories (Manatt, 2013). After the rejection of his first manuscript, Lauterbur was finally able to publish his seminal paper on this new form of imaging in a 1973 publication of *Nature* (Bell et al, 2020). When asked about the rejection, Lauterbur said he did not take it personally (Wade, 2003). He believed that "you could write the entire history of science in the last 50 years in terms of papers rejected by *Science* or *Nature*" (Wade, 2003). In 1985, he joined the faculty at the University of Illinois Urbana-Champaign where he was appointed as the Director of MRI research as well as a professor in the College of Medicine (Bell et al, 2020).

Meanwhile in the United Kingdom, a third key player was also preparing for his entry to the world of magnetic resonance imaging research. After completing his military service, Peter Mansfield attended evening classes at London's Queen Mary College where he was pursuing a degree in Physics (Bell & Murphy et al, 2020). During his studies, he was introduced to Jack Powels, a physicist specializing in Nuclear Magnetic Resonance (Bell & Murphy et al, 2020). Powels went on to supervise Mansfield's undergraduate capstone project which assessed the use of transistor-based spectroscopy to measure the Earth's magnetic fields (Bell & Murphy et al, 2020). After the success of their collaboration, Mansfield was invited to join Powel's NMR research group (Bell & Murphy et al, 2020). It was there that Mansfield successfully defended his Ph.D. thesis "*Proton Magnetic Resonance Relaxation in Solids by Transient Methods*" in 1962 (Bell & Murphy et al, 2020).

DAMADIAN'S INDOMITABLE - THE FIRST FULL-BODY SCANNER

In 1977, Damadian's Indomitable was ready for testing. At its core was a superconducting magnet made of niobium-titanium wire to create a human-sized, hollow cylinder (Wakefield, 2000). On top of it sat a giant liquid helium cooling system that was meant to ensure minimal wire resistance (Wakefield, 2000). However, this system was prone to leaking and resulted in $2000 of helium being consumed each week (University of Washington, 2014). To make matters worse, the helium leaks reduced the strength of the magnet (Wakefield, 2000). Despite the machine's limitations, Damadian and his colleagues were excited to give it a try. Damadian volunteered to be the first subject and attempted a full body scan for hours to no avail (University of Washington, 2014). The team hypothesized that Damadian's high body fat content could be insulating the signals from the wire (University of Washington, 2014). They decided to wait 7 weeks as they kept an eye on Damadian for any concerning side effects. When none were observed, the team re-attempted the scan - this time with Larry Minkoff, a thin graduate student, as their subject. After 5 hours of anxious patience, the team was rewarded with a clear image of the subject's chest (University of Washington, 2014). Motivated by the machine's success, Damadian established his own company, Fonar Corporation, which marketed its first product (based on the original

Indomitable prototype) in 1980 (Wakefield, 2000). In 1985, they received FDA approval for the first mobile MRI scanner (Fonar Corporation, 2021a). Although Damadian's original methods have been rejected due to their inefficiency, Fonar Corporation continues to innovate and produce new MRI technology (Fonar Corporation, 2021b).

MANSFIELD'S LINE SCANNER - A PROMISING COMPETITOR

While conducting research for his Ph.D., Mansfield noticed that after exposure to Nuclear Magnetic Resonance excitation, solid structures emitted an echo (Bell & Murphy et al, 2020). This observation piqued the interest of Charles Slichter at the University of Illinois, who invited Mansfield to work in his lab (Bell & Murphy, 2020). There, Mansfield pursued postgraduate studies with a focus on multi-pulse NMR (Bell & Murphy, 2020). In 1972, he hypothesized that NMR may allow for the study of crystalline solids (Bell & Murphy, 2020). In fact, one particular experiment on this subject initiated the invention of MRI. The experiment involved applying a gradient magnetic field to a sample of camphor that was situated between two pieces of plastic (Bell & Murphy, 2020). In doing so, Mansfield discovered that this technique could measure NMR spectrums, which could then be reconstructed into an image using a Fourier transform (Bell & Murphy, 2020). Around this time, he became aware of Paul Lauterbur who was conducting similar research and was inspired to delve further into investigating the potential of NMR in imaging techniques (Bell & Murphy, 2020).

Upon building a full-sized machine, Mansfield's fellow researchers were too hesitant to volunteer as subjects (Bell & Murphy, 2020). At the time, there was a widespread fear that exposure to magnetic fields could cause myocardial infarction (more commonly known as a heart attack). Eager to reap the fruits of his hard labour, Mansfield bravely volunteered to have a scan performed on his abdomen (Bell & Murphy, 2020). Although the scan produced a clear image, Mansfield was displeased by how long it took to complete (Bell & Murphy, 2020).

SUCCESS AT LAST

In 1977, Mansfield invented echo planar imaging, which drastically changed how images were obtained. This technique relies on rephasing gradients which consist of pulses of echoes presented at different phases (Vadera &

Bashir et al, 2021). In contrast, traditional spin echo sequences rely on re-peated frequencies being presented at pre-defined phases (Vadera & Bashir et al, 2021). Echo planar imaging may rely solely on rephasing gradients or a combination of gradient and spin echo sequences (Vadera & Bashir et al, 2021). This led to the quick, accurate and noninvasive depiction of internal organs (Vogel, 2003). Modern implementations of the echo planar imaging technique allow for images to be produced in 20-100 ms, making it an ideal measure of temporal resolution (Vadera & Bashir et al, 2021). It is commonly used in various types of imaging including, but not limited to, cardiac, ab-dominal, functional, and diffusion imaging (Vadera & Bashir et al, 2021).

Mansfield's first clinical MRI scanner was built in 1980, but it was not made available for clinical use until 1984 (Bell & Murphy, 2020).

THE 2003 NOBEL PRIZE CONTROVERSY

In 2003, the Nobel Prize in Physiology or Medicine was awarded to Paul Lauterbur and Peter Mansfield for their separate contributions to the devel-opment of MRI technology (Partain et al, 2004). Specifically, Lauterbur was being recognized for setting the foundations with gradient imaging and Man-sfield for his development of echo planar imaging (Partain et al, 2004). How-ever, Lauterbur and Mansfield's Nobel Prize win received lots of attention in the scientific community. By the time their contributions were honoured, MRI had become a commonly used tool in both medicine and research (Vogel, 2003). As such, some scientists believed the prize was long overdue. George Radda, an MRI researcher at the University of Oxford argued that their award could have been granted 10 years earlier (Vogel, 2003)! However, opposing critics argue that Mansfield and Lauterbur were not the rightful recipients of the award or that at the very least, they should have been accompanied by Raymond Damadian. In fact, it is believed that the award was delayed for almost a decade due to heavy criticism from rival scientists in the early days of MRI research and implementation (Vogel, 2003).

Among those critics was none other than Damadian himself - and under-standably so. Prior to the Nobel Prize announcement, Damadian had received numerous awards for his contributions to MRI research, both independently and alongside Lauterbur (Breitzman, 2017). Given that the Nobel Prize can be awarded to up to 3 contributors simultaneously, Damadian's omission came as a major shock. It is believed that much of the issue was rooted in personal rivalries between the scientists. Some rumours claimed that Laut-erbur would not accept the Nobel Prize if he were to share it with Damadian

(Breitzman, 2017). Angry, Damadian invested in full-page advertisements in several major newspapers across the US to express his disagreement - a decision that was viewed unfavourably by some members of the scientific community for its lack of professionalism (Partain, 2004). In the advertisements, he claimed that "Lauterbur took his [Damadian's] numerical findings and converted them into pixel intensities to create an image" (Breitzman, 2017), thus making him [Damadian] the rightful recipient of the Nobel Prize.

Naturally, scientists from around the world had much to say regarding this issue. In a 2004 collection of editorials compiled by Leon Partain (MD, PhD), several renowned scientists shared their thoughts on the matter. Among them was William G. Bradley, a medical doctor and radiology professor at the University of California at San Diego, who empathized with Damadian's anger. Bradley noted that the Nobel Prize guidelines allow for up to 3 contributors to share an award, which indicates that Damadian's exclusion was "clearly intentional" (Partain, 2003). He highlights 2 potential reasons why the scientist may have been excluded. The first and less likely hypothesis is that the award focuses specifically on imaging, rather than magnetic resonance in general. It could be argued that Damadian's discovery proved the incidence of prolonged longitudinal relaxation time (T1) in cancer cells, as opposed to establishing an imaging technique. In contrast, Mansfield's contribution (despite coming much later than Damadian's) launched and refined a new imaging technique that was set to become a primary method of diagnosis for many health conditions. Bradely's second, more likely hypothesis is that the judging committee was displeased with Damadian's provocative self-promotion tactics. Several other scientists support this perspective, noting that Damadian had strained relationships with many scientists in his field (Termine & Macchia, 2005). Despite this, Bradley questions the Nobel Committee for failing to look past these behaviours and for not judging participants solely on their scientific contributions. He argues that Damadian should have been awarded for building the Indomitable, the first functional human MRI scanner, just as Sir Godfrey Hounsfield was awarded the Nobel Prize for his invention of the CT scanner.

A 2005 analysis of the situation supported Bradley's stance. Upon analyzing both primary and secondary sources and creating a chronological timeline of key events, Termine and Macchia found that Damadian was indeed the first to propose the concept of a human magnetic resonance scanner as well as the first to receive a patent for it in 1974. Furthermore, Lauterbur referenced Damadian in his 1973 seminal paper discussing the images he had taken of clams. As such, they argued that Damadian's exclusion was indeed intentional and cruel.

Other scientists begged to differ. Ian R. Young (PhD), a member of numerous elite academic societies specializing in radiography, described Damadian's contributions as "lead[ing] precisely nowhere", noting that other scientists' more recent publications investigating NMR and cancer in greater depth held little consensus with the findings of Damadian's work regarding both cancer imaging and diagnosis (Partain, 2004).

Even a decade later, the controversy had yet to subside. In 2012, Morton Meyers reignited the fire with his book *"Prize Fight: The Race and the Rivalry to be the First in Science"*, where he argued that Lauterbur and Mansfield should not have received the 2003 Nobel Prize for the development of MRI, citing Damadian as the rightful recipient based on his 1971 paper (Manatt, 2013). This claim received backlash from several scientists, including the California Institute of Technology's Stan Manatt, for whom Lauterbur had been a consultant in the Jet Propulsion Lab (Manatt, 2013). In his 2013 editorial, Manatt tears apart Meyers' argument. In regards to Damadian's 1971 paper, Manatt states that none of the data "can be extrapolated to show that magnetic field gradients can measure distances and/or concentration gradients of nuclear magnetic resonance (NMR) active nuclei" (2013). Rather, the independent works of Lauterbur and Mansfield clearly highlight these connections, making them more worthy of receiving credit for their development. Furthermore, he argues that while Lauterbur's slow progress could be justified by his struggle to secure funding, Damadian consistently received generous investments from the NIH. Despite this, the latter still failed to make groundbreaking progress. Manatt goes as far as to question Meyers' motives, citing that Lauterbur's attempts to bring MRI research to SUNY Stony Brook's radiology department were blocked by Meyers himself. Regardless, even Manatt's claims should be taken with skepticism due to his personal connections to Lauterbur.

There is no doubt that Lauterbur and Mansfield were not the sole inventors of MRI. As with most major scientific breakthroughs, the invention of MRI scanners was truly a race to the top, with many scientists around the globe scurrying to produce the world's first full-sized scanner. MRI researcher George Radda stands by this but argues that Lauterbur and Mansfield's contributions were miles ahead of their competitors, which (in his eyes) undoubtedly makes them the worthy recipients of the Nobel Prize (Vogel, 2003). Nobel prize or not, Raymond Damadian continues to be considered a pioneer in the field who invented one of the first MRI scanners.

CHAPTER THREE

How Did The Discovery Of MRI Scans Impact Medical Research And Imaging?

Shyla Bhayana

The discovery of MRI scans has had a significant impact on the advancement of the medical field. Since its discovery, MRI technology has become known for its painless, non-invasive diagnostic procedure without radiation, thus posing no potential side effects to the human body (FDA, 2017). MRI scans allow physicians, scientists and researchers to gain valuable information by examining the inside of a patient's body to assess injuries and diagnose a wide variety of diseases and conditions. Along with diagnoses, an MRI scan can also provide important information about how a patient has responded to a specific treatment, which is often useful when monitoring treatment progression (FDA, 2017). As a result of its ability to use a strong magnetic field and radio waves to create detailed images of various organs, systems and tissues, MRI has become a powerful tool used in clinical practice and medical research (Chang et al., 1999).

MRI AND CANCER

Early detection of cancer is vital for a patient's chances of survival, and over the years, MRI technology has played a prominent role in improving cancer diagnosis and research. Unlike other imaging technologies, an MRI scan provides a three-dimensional image with cross-sectional views from various angles. These images help medical professionals make informed decisions about appropriate treatments for the patient, such as surgery and radiation (American Cancer Society, 2019). One of the most significant benefits of using MRI for cancer detection is its ability to produce whole-body images. The versatility of the shape, size and design of MRI machines allows a patient to be scanned from head to toe, which is exceptionally useful for finding various types of cancer occurring anywhere in the body (POM, 2016).

BRAIN AND SPINAL CORD TUMORS

Brain and spinal cord tumors are most commonly diagnosed with MRI scans as they are the best resource to pinpoint the tumor's location in these areas. MRI produces high resolution and extensive anatomic detailed images, providing information about perfusion (blood flow) and tumor cell density (NINDS, 2020). To further improve these images, a contrast dye such as gadolinium can be injected into veins to make details more defined, making the tumor easily identifiable (American Cancer Society, 2020). Special types of MRI have also been developed over time to improve brain and spinal cord tumor diagnoses in certain situations. Magnetic Resonance Spectroscopy (MRS) is a test that is done as part of an MRI scan with the primary purpose of measuring biochemical and metabolic changes in specific areas of the brain. These changes are displayed in a graph-like result called a spectra, and the results from a tumor can be compared to the results from healthy brain tissue to determine what type of tumor is present and how likely it is to grow over time. MRS is also used to determine if any tumor remains after the cancer treatment is completed (American Cancer Society, 2020). Another type of MRI developed to enhance imaging in this area is Magnetic Resonance Perfusion (MRP) or perfusion MRI. After the contrast dye is injected into a vein, a particular type of MRI image is taken to analyze the amount of blood flowing through the tumor and different parts of the brain. These results determine how rapidly the tumor is growing as tumors require a larger blood supply compared to regular areas of the brain. This allows doctors to pick the best place for a biopsy, a medical test that examines a tissue sample to determine the presence or extent of cancer (American Cancer Society, 2020). Lastly, functional MRI (fMRI) is a test that detects blood flow changes to determine the distance between specific brain functions and a tumor. Doctors can use this test to determine which parts of the brain to avoid when performing surgery or radiation therapy (NINDS, 2020).

BREAST CANCER

Ongoing research is being conducted to examine the benefits of using MRI technology for breast cancer screening. Although a mammogram is a more common screening tool for breast cancer, a study conducted by Kriege et al. (2004) suggested that women with a history of breast cancer in their family would benefit more from an MRI scan. This study discovered that MRI technology was more sensitive and detected the tumor better than traditional mammography for women who inherited susceptibility to cancer.

Another study conducted by Warner et al. (2001) discovered that MRI performed on the breast was capable of detecting early stages of breast cancer with approximately 94%-100% sensitivity. Furthermore, out of 196 women who were at high risk of breast cancer due to their inherited susceptibility, MRI technology could detect six stage I invasive cancers and one non-invasive cancer. In contrast, an ultrasound detected three invasive cancers, a mammogram detected only two, and a physical examination detected two. This study provided strong evidence that when detecting breast cancer in women with BRCA1 and BRCA2 mutations (BRCA1 and BRCA2 are genes that indicate an inherited susceptibility to cancer if mutated), mammography or other screening options are less sensitive compared to an MRI.

Hartman et al. (2004) conducted a study in which MRI and mammography were compared for women at high genetic risk for breast carcinoma, the cancer of epithelial cells. They discovered that the breast MRI taken was better able to detect high-grade Ductal Carcinoma In Situ (DCIS), the presence of abnormal cells inside a milk duct and one of the earliest forms of breast cancer. The MRI could also better detect high-risk lesions (abnormalities in tissue), which the mammogram often missed. Additionally, out of forty-one women, three malignant lesions were identified by the MRI, while none were identified by the mammogram, indicating that for some women, MR imaging is a better screening method than mammography.

Over the last ten years, MRI technology has advanced significantly, yielding better image resolution and improved potential for a biopsy (Buchanan et al., 2005). Since medical professionals are more experienced with reading MRIs, specifically breast MRIs, they are now skilled to interpret the scans and make informed decisions with the given information. Additionally, in the past, there have been cases where medical professionals would suspect that a patient's distant metastases (spreading of cancer to other parts of the body) were initially caused by a breast tumor. In these cases, if a physical exam or mammogram does not indicate where the tumor is present, the patient may have to proceed with a mastectomy, the surgical removal of the breasts. With high resolution images produced by an MRI, tumors can be found more easily, and a mastectomy can be avoided.

CERVICAL CANCER

Cervical cancer, the leading cause of mortality for women globally, is caused by abnormal cells located in the cervix or by a previous infection from the human papillomavirus (HPV) (RSNA, 2018). For many years, the traditional examination methods have been based on clinical exams and basic imaging such as intravenous pyelogram and x-ray. Despite advances in radiotherapy and chemotherapy, local control of cervical cancer has been insufficient in the past. Recently, this situation has changed due to the increased implementation of MRI-guided brachytherapy (a type of radiation therapy used to treat cancer) (Fields & Weiss, 2016).

Radiation therapy such as brachytherapy is a challenging treatment option for cervical cancer as it requires high radiation doses to achieve tumor control. Additionally, the tumor is often positioned directly between radiosensitive organs such as the bladder and bowel, increasing the risk of using high doses of radiation therapy to target the tumor (Fields & Weiss, 2016). To address this issue, MRI technology has been implemented in many aspects of tumor staging, planning and delivery of radiotherapy, as well as post-treatment surveillance (Paulson et al., 2015). Since MRI is well known for its exceptional soft-tissue imaging characteristics, it provides better results than clinical exams and other 3D imaging techniques mentioned earlier (Hricak et al., 2007). MRI also provides medical professionals with a clear and precise visualization of the cervical tumor, allowing for a more reliable, volumetric definition of the target volume for brachytherapy planning (Paulson et al., 2015). With the addition of MRI to brachytherapy planning, tumor control rates have significantly improved (Pötter et al., 2011).

MRI technology has been applied clinically to help medical professionals determine the need for additional surgical treatment in the presence of residual tumor and to detect possible tumor recurrence post-treatment (Hatano et al., 1999). MRI technology has exceptional accuracy in detecting residual tumors for both central areas such as the vagina and cervix as well as in the parametria and pelvic side walls, which are often difficult to detect with just a clinical exam.

CARDIOVASCULAR MRI

Over the years, cardiovascular MRI has evolved from a helpful research tool into a clinically proven, safe, and comprehensive imaging method (Marcu et al., 2006). Cardiovascular MRI is optimized for usage in the cardiovascular system by producing detailed images of structures within and surrounding the heart. These precise images provide anatomic and functional information about the patient's acquired or congenital (since birth) heart disease and can help assess heart defects or heart failures (Marcu et al., 2006). Since cardiovascular MRI produces detailed images without the use of radiation, it is considered to be one of the best imaging tools for heart conditions (Krans, 2018).

By examining a cardiovascular MRI scan, a medical professional can evaluate the function of a patient's heart chambers, heart valves, major vessels (both size and blood flow), and the surrounding structures such as the pericardium (a sac that surrounds the heart) (RSNA, 2018). Using this information, they are then able to diagnose various cardiovascular disorders such as heart tumors, coronary artery diseases and pericardial diseases (a disease that affects the outer lining of the heart). Additionally, medical professionals can use MRI scans to plan a patient's treatment and monitor their progression over time once a diagnosis is made.

REPLACING BIOPSIES WITH SOUND

Chronic liver disease and cirrhosis are a large source of morbidity and mortality globally (Sepanlou et al., 2020). The National Institute of Biomedical Imaging and Bioengineering (NIBIB) funded research in which scientists have been developing a method to turn sound waves into images of the liver (NIBIB, 2018). This technique provides a new pain-free and non-invasive approach to find tumors or damage in tissue caused by liver disease. A device called magnetic resonance elastography (MRE) is placed on the patient's liver before they enter the MRI machine. This device then pulses sound waves through the liver, which the MRI machine is able to detect and use to analyze the health and density of the patient's liver tissue. Compared to a traditional biopsy, this method is safer, less expensive and is often more comfortable for the patient. Since MREs are advancing and are able to recognize minimal differences in tissue density, there is a strong possibility that they could be used to detect cancer in the near future.

MRI FOR CHILDREN

MRI is considered one of the best imaging tools for children as, unlike a CT scan, it does not have any harmful ionizing radiation (NIBIB, 2018). However, a common challenge that MRI technicians face is acquiring a clear scan for younger patients or for patients unable to stay still for long periods of time. Subsequently, many young children are required to undergo anesthesia which increases the health risks for the patient. NIBIB funded research to develop a robust pediatric MRI. By developing a pediatric coil that is specifically designed for smaller bodies, an image can be produced clearly, quickly and will demand less MRI operator skill. This design has the potential to provide a cheaper and safer option for children.

Another project funded by the NIBIB is also trying to resolve this issue from a different approach. Researchers are developing a motion correction system that has the ability to improve the quality of MRI exams (NIBIB, 2018). Additionally, they are developing an optical tracking system that can adapt and match MRI pulses to changes in the patient's pose in real-time. This system has the potential to reduce the cost of repeating MRI exams due to their lack of quality and reduce the amount of anesthesia used to perform an MRI exam.

DETERMINING CHARACTERISTICS OF A TUMOR

Unlike PET or SPECT scans, a traditional MRI scan is unable to measure metabolic rates (NIBIB, 2018). Researchers funded by the NIBIB have developed a technique for prostate cancer patients which allows medical professionals to inject specific specialized compounds such as hyperpolarized carbon-13 into the cancer patient to better analyze and measure the metabolic rate of the tumor. The information gained from this technique can provide accurate insight regarding a tumor's aggressiveness. Monitoring the progression of tumor aggressiveness can help medical professionals make educated predictions about the risks involved with the patient's specific type of cancer. This information is essential for prostate cancer patients who are often forced to follow a "wait and watch" approach filled with uncertainty.

IS THE OUTCOME WORTH THE COST?

The concept of value has a different significance depending on the situation or person. Oftentimes, we can denote value as "outcome over cost," which, in this case, represents the accurate information and guidance that imaging technology such as MRI provides. It is well recognized that MRI is a non-invasive and safe imaging test that has optimized patient care by playing a prominent role in diagnosis, treatment and patient prognosis (Van Beek et al., 2018). However, the cost of purchasing, delivering and maintaining an MRI machine is relatively high (over 3 million dollars per machine) and is something that needs to be considered (Reed, 2019). This leads us to the main question: is the outcome worth the cost? To answer this question, various aspects need to be considered in the discussion of MRI value:

1. MRI scans are often perceived negatively in a patient's mind as they are very costly and are taken for unfortunate reasons (injuries, disease diagnosis, etc.). Thus, MRI technology needs to be showcased more appealingly in order to increase the patient's perception of value. To improve patient experiences, more comfortable and shorter scans with less noise need to be implemented.

2. An MRI machine's report time increases with scan complexity. By optimizing protocols, reducing the number of scans required per patient and developing new artificial intelligence to aid the process, MRI technology can improve exam read time and, thus, its overall value.

3 In certain situations, such as a research setting, areas where constraints of economic resources are less problematic, and when a complex MRI is the only diagnostic modality, MRI is found to be a very useful tool. However, in many other settings, a highly detailed MRI without subsequent appropriate treatment may seem invaluable when compared to its cost.

Overall, the question posed earlier has no right answer as various factors must be taken into account when discussing MRI value. Depending on what the MRI is being used for, and for the patients who are using it, MRI value is debatable.

CONCLUSION

This chapter discussed the various applications of MRI technology in both a clinical and research setting. As MRI technology becomes more prominent in the medical field and experience with it accumulates, MRI seems to be a very capable screening tool with many benefits. MRI scans have not only allowed medical professionals, scientists and researchers to obtain useful information but have also improved early detection and diagnosis, radiation therapy, treatments and patient prognosis significantly. MRI technology continues to evolve, introducing new scanning methods that continue to positively impact diagnostic imaging standards of practice and developing new systems and designs that produce better image quality and yield more significant throughput.

CHAPTER FOUR

What Are MRI Scans?

Simarjit Kaur Sidhu

Magnetic resonance Imaging (MRI) is an imaging technique used to produce cross-sectional images of various organs and tissues of our body with the use of a large magnetic field and radiofrequency waves. MRI scans are used for both diagnostic and therapeutic purposes. Major use of MRI is in the field of neuroradiology and musculoskeletal radiology (Kissane et al., 2020, 33-34). The basic concept of magnetic resonance involves the process of energy transfer. When a patient is exposed to a certain magnitude of energy, some of it is absorbed that is reemitted later. This reemitted energy then can be detected by specific detectors and information is processed by computers and special softwares to decode into images (Brown & Semelka, 2003, 11-12). The basic working principles of magnetic resonance have been known for more than 70 years and many significant advancements have been achieved. Benefits that have been achieved are improved speed of image acquisition, enhanced quality and definition of images, better information extraction. There has been incredibly sophisticated hardware development that has helped to achieve above benefits (Wald, 2019, 139-144).

WORKING PRINCIPLE

MRI technology is based on the principle of electromagnetic effect of rotating protons and hydrogen atoms in water and other body matter. MRI produces data sets that can be reconstructed as two-dimensional cross-sectional representations or three-dimensional volumes of anatomic structures with excellent soft-tissue contrast using a magnetic field and high frequency electromagnetic pulses (Biederer, 2005, 62-72). To perform a MRI scan, a patient is laid in a machine that produces a strong uniform magnetic field. This field results in alignment of hydrogen nuclei inside the patient's body in the direction of the magnetic field. An external radiofrequency pulse (RF) disturbs these nuclei from their current orientation. Following the termination of the RF pulse, the hydrogen nuclei revert back to their original alignment within the externally applied magnetic field. This shift results in a signal due to loss of energy which is then processed by a computer to generate an image. The frequency of the signal depends on the strength of the external field. The exact location of the hydrogen molecule can be calculated and assigned a colour from white to black on the detector resulting in the final image or scan (Kissane et al., 2020, 33-36). This topic is discussed in more detail in Chapter 6.

TYPES OF MRI SCANS

MRI scans can be broadly classified as structural and functional scans. Structural MRI scans provide static anatomical information of the region being imaged (Symms et al., n.d., 1234-1244).

There are two main sequences or types of structural MRI, T1- weighted and T2- weighted sequences. These are based on special properties of hydrogen atoms when exposed to the magnetic field. These sequences differ how water and fat appear on the scan. In T1- weighted scan, water appears as dark or hypointense, fat appears as bright or hyperintense. Whereas T2- weighted scans are opposite where water appears dark or hyperintense and fat as dark or hypointense. Other soft tissues appear as a gradient between dark and bright for both the sequences. There are other complicated forms of MRI like gradient, Fluid Attenuated inversion recovery (FLAIR), echo which are based on these T1 and T2 weighted sequences (Kissane et al., 2020, 33-36).

Major use of structural MRI is in looking for any anatomical abnormalities in the body. T1 weighted scans are used widely for research purposes like in Alzheimer Disease research to study hippocampus. The use for clinical purpose is in diagnosis of epilepsy, abnormalities of cranial nerves. T2 weighted scans are more useful for research studies on Parkinson's disease where Iron content in the brain is studied. T2 weighted scans are very useful in the clinic to diagnose bleeds like acute cerebral hemorrhage, Subarachnoid hemorrhage, cerebral microhemorrhages (Symms et al., n.d., 1234-1244).

Functional MRI (fMRI) scanning is the other type of MRI imaging that provides dynamic physiological information. Various types of fMRI include Blood Oxygen level dependent (BOLD) technique, perfusion studies, blood flow studies, cerebrospinal fluid pulsation measurements, phase contrast flow measurement, magnetic resonance angiography, magnetisation transfer imaging, functional brain imaging, noble gas imaging, ultra-high field imaging. Research widely using magnetic transfer imaging scans for studies looking into various disorders like schizophrenia, dementia, traumatic brain injury, brain tumors, multiple sclerosis and many more. Diffusion imaging is used in research concerning epilepsy, dementia, trauma, multiple sclerosis, to study postoperative sequelae, amyotrophic lateral sclerosis, schizophrenia, studies on normal brain maturation and aging. The clinic uses diffusion studies for diagnostic purposes of infraction, tumors, Creutzfeldt-Jakob disease (CJD) (Symms et al., n.d., 1234-1244).

Magnetic resonance angiography (MRA) is an advanced form of MRI that is used to investigate the flow of blood in the vascular system. It is used to see the flow is laminar or disturbed, if the patency of the blood vessels is normal or not. Stenosis or any other kinds of obstructions that may disrupt the normal laminar flow of blood in the vascular network can easily be seen through MR angiography. This technique can also be done without the use of contrast dyes that gives an additional advantage of being able to perform multiple scans. The "bright-blood" image sequence is the most widely used sequence. The other technique of "black-blood" MRA is seldom used these days. Combining advanced techniques like gradient echo sequences, saturation and presaturation pulses help in generating advanced data in 2-D and 3-D. Various other modes of MRA include time-of- flight MRA, phase contrast MRA (Brown & Semelka, 2003,151-164).

Magnetic resonance spectroscopy (MRS) is the technique that is used to study individual molecules within a sample. Magnetic resonance was initially developed for this technique. With the development of whole body scanners, it has become easier to study the biochemistry of various diseases and disorders (Brown & Semelka, 2003, 181-196).

MRI imaging has various shortcomings like motion artefacts that can compromise the quality of scans. There are various techniques that are used to reduce these artefacts, and aid in better detection of a lesion in the specific organ or region of the body. These techniques include fat suppression, water enhancement, pulse sequences like spin echo, gradient echo, contrast studies, magnetization- transfer suppression, triggering/gating and many more. (Brown & Semelka, 2003) The use of various such techniques and various combinations depends on the purpose of MRI and the region which is being imaged. The advances in the hardware technology of magnetic resonance imaging have been remarkable. This has aided in generation of more sophisticated images.

INDICATIONS

MRI is indicated for diagnostic and therapeutic purposes. MRI scan produces accurate and reproducible imaging of the pathology leading to good visualization of disease process. The protocol for MRI scanning according to various anatomical regions of body are as below (Brown & Semelka, 2003, 223-246).

1. Brain and Spine- Diffusion weighted studies are used for visualization infarction, hemorrhages and strokes. FLAIR is used to study white matter inflammation. Magnetic resonance angiography is used for studying cerebrovascular disease. T1 and T2 weighted imaging is used to look for structural abnormalities and tumors. Sagittal and transverse sections are used for spine to investigate disc herniations, nerve root compressions. Contrast based studies are used routinely to investigate neoplasms of the region.

2. Neck and thorax- T1 and T2 weighted sequences are done if MRI is being done to look for structural abnormalities in neck, chest and mediastinum. Gadolinium based contrast studies are used to investigate neoplastic changes in thyroid, parathyroid, lymph nodes and other organs of the region. Holding of the breath while taking images improves visualization of lesions of lung periphery. When investigat-

3 ing breast tissue, techniques that reduce fat signals are employed to enhance the quality of scans.

Musculoskeletal- MRI is routinely used to investigate various disorders of musculoskeletal system like dislocations, avascular necrosis, traumatic injuries. Again contrast based studies are more useful when investigating neoplastic or inflammatory diseases.

4 Vascular system and heart- Magnetic resonance angiography is widely used to study aorta and other great vessels of the body. MRI is also used to diagnose congenital heart defects. T1 weights echo is widely used for this.

5 Liver and other abdominal organs- T1 and T2 weighted images are commonly used. Contrast based studies help in better visualization of lesions. Various contrast agents that are used are gadolinium chelated agents, MnDPDP, Feridex. Breath holding images also help to enhance the quality. Bowels are better distinguished from other structures using ETSE (echo train spin echo) technique.

6 Pelvis- Sagittal sections are particularly useful for the pelvic region. Using Gadolinium chelated contrast agents help enhance the visualization of pelvic and adnexal masses. When MRI is done due to clinical suspicion of ovarian cancer in females, use of fat suppression techniques can lead to high yield results.

CONTRAINDICATIONS

There are several patient related factors that make them unsuitable to under go a MRI scan. A patient who has a metalling implant like a cardiac pacemaker, aneurysm clip, ear implants, a metallic foreign body in their eyes cannot undergo a MRI scan even if their symptoms make them a candidate that needs a scan to aid in their diagnosis and treatment.

Patients who have a severe form of claustrophobia and cannot tolerate lying in the tube shaped MRI machine are other group of patients where MRI cannot be done easily. Sedatives might be used in some to facilitate the process (Ullrich, n.d.).

Use of contrast dyes is contraindicated in patients who have a history of allergic reaction to the dye that was used in the past. MRI without contrast studies can still be performed safely in this group.

ADVANTAGES

MRI being a non-invasive technique is very prevalent these days. It has the major advantage of not using ionizing radiations making it a very safer imaging technique than CT scan. Anatomy of soft tissues in our body is displayed on the MRI scan making it very useful in diagnosis and treatment of disorders of brain, spine, joints, muscles, tendons, ligaments, bone and soft tissue tumors. The unique aspect of MRI is that it is able to image structural as well as functional aspects of the organs. The amount of information produced in one field of view of a MRI scan can be incredibly huge due to the ability of different magnetic fields sequences to produce different displays of various tissues (Kissane et al., 2020, 33-36).

DISADVANTAGES

MRI scanning is a lengthy process as compared to techniques like CT scan. This shortcoming leads it to not being an imaging technique of choice in emergency situations. It is also a costlier technique than CT scan and is not much readily available in all facilities.
Patients have to stay still while the imaging is being performed as any motion leads to blurry images. This can be cumbersome for the patients. Another set of patients who have claustrophobia or anxiety, pediatric patients are not very comfortable inside the tube shaped MRI machine. They sometimes have to be sedated in order to get good images.

Another disadvantage of MRI is that it cannot be done in patients who have implants like a cardiac pacemaker, an ear implant. Due to the ferromagnetic properties of the material used to manufacture these devices, they can move and cause discomfort to patients in the strong magnetic field. This leads to inability to form a good diagnosis due to inability to perform an absolutely indicated imaging technique.

The contrast dye that is used for more detailed scanning has its own set of side effects and risks. It can sometimes lead to allergic reactions. Although this is very rare, it can be fatal if anaphylaxis happens due to the dye. The dye can cause harm to kidneys leading to fibrosis, this can sometimes be irreversible.

Having proper patient information and history of the patient can minimise the risks of these side effects (Kissane et al., 2020, 34-35).

WHAT TO EXPECT WHEN HAVING A MRI SCAN DONE?

Patients are advised to get a MRI scan done in order to aid the physicians in building a proper diagnosis and care plan. Depending on which part of the body needs to be scanned, there can be some preparations involved like fasting for a few hours prior to a scan or drinking a fairly large amount of fluids. Once the patient is at the scanning facility, patient is provided a form where they have to answer a few questions regarding their current and past health conditions, the medications they are on, any allergies to contrasts, any implants or pieces of jewellery that cannot be taken out, previous history of surgeries, history of claustrophobia or anxiety. These questionnaire are basically designed to rule out any contraindication to MRI and to be prepared in case an emergency situation arises. Next the patient is made ready to go inside the MRI room by changing into a hospital gown and ensuring all metals or jewellery are removed from the body. If the MRI scan involves use of a contrast then an IV line is set up through which contrast dye is delivered. After this, the patient lies in the tube shaped MRI machine and technician ensures proper positioning and provides appropriate guidelines like providing an emergency stop button in case the patient faces any difficulties, providing ear protection to the patient as the MRI machine produces very loud thumping and bumping noises, providing special breathing instructions and so on. Patients who suffer from claustrophobia or pediatric patients may be sedated in order to facilitate the imaging process. Patient has to lie still during the imaging process as movement leads to blurry images.

Time taken to complete the MRI scan depends on which region is being scanned, if the patient is able to follow breathing or any special instructions as directed. It can be anywhere between 15 minutes to an hour. Following the scan, the IV line is removed. The patient is instructed to drink plenty of fluids on the following days in order to flush out the contrast dye rapidly. It is made sure by the staff that the patient is doing alright before being discharged from the facility.

The MRI scan is then reviewed by a radiologist. A comprehensive report is sent to the requesting physician who discusses it with the patient.

CONCLUSION

There has been significant medical advancements since the advent of MRI technique. Diseases can be better diagnosed and at earlier stages. This ultimately leads to better treatment and prognosis. The accessibility and non-invasive nature of MRI makes it a very useful imaging technique. Further chapters of this book discuss the MRI machine and its working, various advancements in the field of magnetic resonance imaging, comparing it with other sophisticated imaging methods.

CHAPTER FIVE

Why Are MRI Scans Worth Investigating?

Simran Bakshi

Despite being a revolutionary and versatile diagnostic tool in modern medicine, MRI is still considered cost-intensive and a major contributor of healthcare expenses (van Beek et al., 2019). This chapter discusses the value of MRI scans in medical diagnoses and its superiority relative to other contemporary imaging technologies. Greater understanding of the clinical applications of MRI may allow imaging researchers to broaden the horizons of MRI use within the medical field. In addition, further investigation of MRI technologies can accelerate advancements in its hardware and enhance its efficacy in the near future.

THE USE OF MRI IN VARIOUS CLINICAL APPLICATIONS

MRI scans have a promising future in the medical research industry and are a powerful clinical tool that can diagnose various clinical diseases. This section will highlight the application of MRI in the diagnosis and treatment planning of endometriosis, autism spectrum disorder, breast cancer, fetal diseases, prostate cancer and multiple sclerosis. In addition, this section will highlight the areas of focus researchers should highlight when investigating MRI technologies. The investigation of MRI scans is crucial to enhance the efficacy and efficiency of imaging technologies that can support further diagnosis and treatment procedures.

ENDOMETRIOSIS

Endometriosis is a chronic gynecologic disease where the endometrium and stroma develop on the outside of the uterine cavity. It affects up to 10% of premenopausal women and causes infertility and pelvic pain (Siegelman & Oliver, 2012). In most instances, laparoscopic surgery is performed by surgeons to remove excess tissue. However, when the endometrium attaches to other organs (e.g., ovaries, fallopian tubes), a more invasive surgery is required. Hence, MRI scanners are used as a diagnostic tool prior to surgery in order to support clinical management, enhance surgical planning and optimize patient care (Bartlett et al., 2019). Consequently, surgeons can scope and decide whether patients with endometriosis must undergo high-risk, invasive surgeries that involve larger abdominal incisions.

Multiple studies have found that MRI is more reliable when diagnosing endometriosis compared to ultrasound (Bartlett et al., 2019; Ibrahim et al., 2012; Tadros & Keriakos, 2016). Pelvic MR imaging detected 61% of patients with endometriosis whereas ultrasound was only able to detect 22% of patients with endometriosis (Bartlett et al., 2019). Interestingly, 51% of patients who initially had a negative pelvic ultrasound had a positive pelvic MR examination (Bartlett et al., 2019). Thus, MRI imaging has the ability to combat false negative results from pelvic ultrasounds. Another study conducted by Hsu et al. (2011) found that MRI is more effective in diagnosing rectosigmoid lesions and endometriosis of the bladder. Correspondingly, MRI has a diagnostic accuracy of approximately 98%, specificity up to 90% and sensitivity of 88% for the diagnosis of bladder endometriosis (Foti et al., 2018). The powerful nature of MR imaging is attributed to its higher contrast resolution and broader field of view relative to ultrasound technologies (Bourgioti et al., 2017). In addition, MR imaging is advantageous relative to other diagnostic imaging as it permits a complete and simultaneous assessment of both the anterior and posterior pelvic compartments (Bianek-Bodzak, 2013). Also, image acquisition is more reproducible in MRIs and data collected from various sequences permits a greater specificity in the characterization of ovarian lesions (Jha et al., 2019). Hence, further research pertaining to the beneficial use of MRI within endometriosis patients can improve patient outcomes.

Recently, 'real-time virtual sonography' (RVS), has been implemented in the diagnosis and treatment of tumours and prenatal imaging. RVS involves the use of computer software to produce a synchronized display of real-time ultrasound images and multiplanar reconstruction images from MRI that correspond to the image plane of the real-time ultrasound (Millischer et al., 2015). RVS has the potential to improve the visualization of the major pelvic anatomical sites (Millischer et al., 2015). Thus, the complementary nature of MRI with transvaginal sonography (TVS) allows for the combination of accuracy and tissue contrast abilities from MRI to amalgamate with the dynamic, real-time qualities of TVS.

AUTISM SPECTRUM DISORDER

Autism spectrum disorder (ASD) is a neurodevelopmental syndrome that is associated with repetitive behaviours and impaired social communication abilities that emerge at 24 months of age (Anagnostou & Jaylor, 2011; McAlonan et al., 2005). ASD causes many financial, societal, educational and health-related hardships for patients and their families (Williamson et al., 2020). Hence, early behavioural interventions can lead to improved outcomes for ASD patients and reduced ASD-related costs. MRI is increasingly being used as a diagnostic tool by neuroradiology researchers to examine ASD within patients. Compared to computerized tomography (CT) or positron emission tomography (PET) scanning, MRI does not involve invasive procedures or injection of radioactive isotopes, respectively (Gutierrez, 2011). The lack of radiation exposure in MRI is particularly vital for children and adolescents (Chen et al., 2011). Structural MRI is capable of providing information pertaining to brain anatomy by detecting grey- and white-matter volumes (Dichter, 2012). A study by Anagnostou and Jaylor (2011) reviewed structural and volumetric MRI data for ASD and found a 10% increase in brain volume that peaks at approximately 2 to 4 years of age. In addition, two studies found that the the use of MRI was able to identify, with an accuracy of 81.0% to 81.8%, whether high-risk siblings of children with autism will receive a clinical diagnosis of ASD at 24 months of age (Emerson et al., 2017; Hazlett et al., 2017). Another study conducted by Davidovitch et al. (1996) observed a significantly larger head circumference within children with autism. Thus, MRI is a possible cost-effective imaging technique that allows for presymptomatic identification of ASD among high-risk children (Williamson et al., 2020).

A specific type of MRI known as 'functional MRI' (fMRI), is a technique that assesses brain activity by identifying changes in blood oxygenation associated with neural activity when an individual is performing a cognitive task (Glover, 2011). Correspondingly, a study that used fMRI by Gervais et al. (2004) found that individuals with autism demonstrated a normal activation pattern when responding to nonvocal sounds however they did not activate superior temporal sulcus (STS) voice-selective regions in response to vocal sounds. In addition, multiple studies used fMRI and found differential lateralization patterns in ASDs, specifically decreased activity in the left hemisphere of the brain (Kleinhans et al., 2008; Müller et al., 1999; Redcay & Courchesne, 2008). Another study conducted by Piven et al. (1997) used T1-weighted MRI to assess the size of various subregions of the corpus callosum in individuals with autism. They observed a significantly decreased size of the body and posterior subregions of the corpus callosum in individuals with autism compared to the healthy comparison subjects. According to Bullmore (2012), fMRI will increasingly be implemented as a phenotypic marker of neurological systems in order to support our understanding of the functional responses of genetic variations (e.g., single nucleotide polymorphisms (SNPs), copy number variation (CNVs), haplotypes). Therefore, MRI has the ability to both, structurally and functionally, analyze physiological activities and has the capacity to advance in complexity in the near future.

BREAST CANCER

With recent advancements in coil technology and development of contrast agents, MRI is playing an increasingly vital role in the detection and management of breast cancer patients. MRI can provide information about invasive lobular cancer, women with dense breasts and the development of large tumours (Lalonde et al., 2005). This information is significantly valuable to surgical planning and has the potential to decrease reoperation rates. According to Mann et al. (2019), breast MRI has the highest sensitivity for breast cancer detection compared to other contemporary imaging modalities. However, the major drawback of MRI is its low to moderate levels of specificity, which in combination with its high sensitivity, may lead to higher patient anxiety levels, lower cost-efficacy and performance of unneeded surgical procedures (Morris & Liberman, 2015). The sensitivity and specificity of MRI in the contralateral breast was found to be 91% and 88%, respectively (Lehman et al., 2007). MRI is also found to be superior to mammography where in a study conducted by Raikhlin et al. (2015), 92.3% of total cancers were identified by

MRI whereas only 30.8% were identified by mammography. In addition, according to a multi-analysis conducted by Warner et al. (2009), the sensitivity of MRI was higher (64-100%) relative to mammography (32-40%). Recently, a new idea known as the abbreviated breast MRI protocol has gained attention in the clinical field. This protocol is based on the original, full diagnostic protocol and entails an unenhanced, a single contrast-enhanced, a generated subtraction, and a single maximum intensity projection (MIP) image (Chhor & Mercado, 2017). With this adjusted protocol, clinicians are able to reduce the MRI acquisition time from 17 minutes to 3 minutes (Kuhl et al., 2014). Noteworthily, they found that the cancer yield and diagnostic accuracy were not significantly hindered by the shift from a full diagnostic protocol to an abbreviated protocol. Hence, further investigation into altered forms of MRI, such as the abbreviated protocol, can aid in the efficiency of clinical technological practices without disrupting patient outcomes or cancer yield.

FETAL MRI

Despite ultrasound being the golden standard of fetal assessment, the use of fetal MRI has been gaining traction for evaluating pregnancy-related disorders over the past 20 years. According to Saleem (2014), visualization conducted by fetal MRI is not significantly restricted by certain characteristics, such as maternal obesity, fetal position, or oligohydramnios. Oligohydramnios refers to a decreased amniotic fluid volume for gestational age (Keilman & Shanks, 2021). Compared to prenatal sonography, fetal MR imaging has enhanced contrast resolution and allows for visualization of both the posterior and anterior cerebral hemisphere of the fetal brain (Glenn & Barkovich, 2006). This is beneficial as prenatal sonography often involves the anterior cerebral hemisphere to be shadowed and results in an uneven visualization of the brain.

Fetal MRI is predominantly conducted to assess a suspected abnormality identified by prenatal sonography. A study conducted by Levine et al. (1999) found that MRI findings caused a change in diagnoses of 40% fetuses with abnormal confirmatory sonograms. Hence, fetal MRI is capable of providing additional information about abnormalities that may aid in prenatal counseling in addition to counselling of the recurrence risk of future pregnancies (Glenn, 2010). Currently, there has been an increasing amount of research for new possibilities, such as performing fetal MRI at 3.0 teslas (T) compared to its usual field strength of 1.5 T. This interest in the enhancement of field strengths is driven by a possibility of increasing signal-to-noise ratio (SNR), fetal depiction precision, and spatial resolution (Weisstanner et al., 2017). In addition, many institutions who solely have access to 3.0 T MRI can also perform fetal MRI without any restrictions. Hence, additional research about the application of various MRI technologies can allow clinicians to harness the power of certain MRI components (e.g., field strength of magnets) and increase the effectiveness of diagnosis protocols.

PROSTATE CANCER

Since its introduction in the clinical field, MRI has played a vital role in the diagnosis of prostate cancer. According to Ahmed et al. (2009), transrectal ultrasound (TRUS)-guided prostate biopsy has a negative predictive value (NPV) of 70-80%. Therefore, approximately 20-30% of patients with a negative biopsy still have the possibility of being diagnosed with prostate cancer. A study conducted by Murphy et al. (2013) found that multiparametric MRI can combat diagnostic barriers, such as false-negative biopsies, by detecting tumours missed by biopsy. These missed tumours are predominantly located deep in the prostate and away from the rectal wall where the TRUS biopsy can reach. A study by Park et al. (2011) used 3-T MRI found higher cancer detection (29.5% vs 9.8%) and a significantly higher positive core rate compared to the non-MRI group (10.0% vs 2.5%). Hence, MRI performed before repeat biopsy is a potential procedure that can be implemented for patients with a history of negative biopsy and consistently high prostate-specific antigen (PSA) levels as MR imaging can indicate the location of the cancer being investigated. In addition, Bonekamp et al. (2011) states that MRI-guided prostate biopsy allows for higher precision when targeting. MRI-guided prostate biopsy entails the combination of ultrasound and MRI or solely MRI. Two studies have concluded that the fusion of ultrasound-guided and MRI-guided prostate biopsy have a higher potential of detecting cancer per core than standard 12-core TRUS (Natarajan et al., 2011; Pinto et al., 2011). Prostate MRI also plays a role in treatment planning and can aid in the stratification of patients into the optimal therapy option. For example, preoperative MRI can detect extracapsular extension (ECE) and seminal vesicle invasion (SVI) which can prevent curative surgery as only carcinomas located within the prostate gland are curable by radical prostatectomy (RP) (Murphy et al., 2013). Preoperative MRI has demonstrated high sensitivity (0.78) and specificity (0.98) in the assessment of ECE and SVI, respectively (Porcaro et al., 2013). In addition, MRI can detect a lack of disruption of the NVB, hence allowing the surgeon to consider a NVB-sparing surgery (Carroll et al., 2006). Therefore, preoperative prostate MRI has the potential to deter from unnecessary surgical procedures and optimize patient health.

MULTIPLE SCLEROSIS

Pathological evaluation is considered to be the ideal intervention when ex-amining disease manifestations and treatment planning related to multiple sclerosis (MS). However, as a result of the high specificity of MRI, it has been playing an increasing role as a possible clinical tool to detect abnormalities pertaining to MS. For example, Meier et al. (2007) implemented a time-series model to evaluate dynamic signal variation associated with lesion dynamics. Through the use of MRI, they were able to discover a correlation between lesion dynamics and rates of atrophy and disability. Hence, variation in lesion dynamics may signify a shift from an inflammatory to degenerative pathol-ogy (Meier et al., 2007). Interestingly, a study conducted by Tomassini et al. (2005) used conventional MRI to investigate the correlation between serum sex hormone concentrations and indicators of tissue damage. Through the application of MRI, they were able to conclude that oestrogens and testoster-one play a role in the inflammation, damage, and repair mechanism of MS. Another study conducted by Agosta et al. (2008) used a spinal cord spin-echo based fMRI to identify various motor and sensory paradigms within relaps-ing MS patients. They found that patients demonstrated an altered pattern of functional activation in the cervical cord during a flexion-extension of the right upper limb. In addition, increased activation of the cervical cord was detected in patients with relapsing MS which was correlated with cord struc-tural damage (Gass et al., 2015). Further investigation of MRI and technolog-ical advancements will allow us to broaden the confines of the use of MRI in pathologic disease mechanisms and improve patient outcomes.

WHAT SHOULD MRI RESEARCHERS SHIFT THEIR FOCUS TO?

VALUE-BASED CARE ANALYSIS

Medical imaging, particularly MRI, has revolutionized the delivery of patient care and the methods healthcare providers use to diagnose diseases, assess severity, and examine biological mechanisms of disease physiology. However, when the value-based care model is considered, we must ask whether MRI is capable of changing treatment decisions, enhancing patient outcomes, and improving cost-effectiveness. Anazi et al. (2019) believe that research and innovation is a fundamental pillar that supports the strategy of enhancing MRI for main stakeholders — patients, referring physicians and care teams, hospitals and health institutions, and purchasers. Within the past 10 years,

MRI has advanced to a great extent with enhanced image quality and spatial resolution (Harisinghani et al., 2019). They eliminate the undertaking of invasive procedures, such as biopsies or surgeries, and serve as noninvasive biomarkers (Anazi et al., 2019).

For example, a type of MRI called diffusion-weighted imaging (DWI), is capable of visualizing breast lesions and distinguishing benign from malignant findings (Baltzer et al., 2019). In addition, DWI maps can portray changes in the brain as early as 30 minutes after the onset of a stroke (Chilla et al., 2015). Other advancements in MRI technology, such as MR fat quantification and MR elastography, can identify significant steatosis or fibrosis in liver donor candidates with high sensitivity (Yoon et al., 2015). Hence, scientific discoveries in the diagnostic imaging field have allowed for a deviation from invasive procedures and additional patient complications.

While various MRI researchers focus on technological developments, it is crucial to shift our investigation efforts to technological assessment. Fryback and Thornbury (1991) developed The Technology Assessment Hierarchy model that constructed a framework to assess imaging efficacy. It emphasizes the importance of conducting imaging research beyond the confines of diagnostic accuracy and other image interpretation parameters (e.g., specificity, sensitivity). Anazi et al. (2019) state that imaging researchers should target Levels 3 to 6 of the model. Levels 1 and 2 pertain to the technical quality of the images and diagnostic accuracy, sensitivity, and specificity, respectively. Level 3, called "diagnostic impact", refers to whether information acquired from the imaging technology facilitates a change in the referring physician's diagnosis. Level 4, known as "therapeutic impact", pertains to the impact of medical imaging on a patient's treatment plan. Level 5 refers to the computation of the effect of imaging information on patient outcomes. Lastly, Level 6 assesses the societal costs and benefits of a particular diagnostic imaging modality. Hence, further investigation of MRI scans is required to ensure patient outcomes, treatment decisions and cost-effectiveness is prioritized when analyzing the impact of diagnosing image technologies.

A PROMISING FUTURE

MRI is not merely a clinical tool, but an imaging technology that can act as a catalyst for the imaging modality industry as a whole. Further investigation, research and development of MRI can increase the prominence of non-invasive procedures and ensure positive patient outcomes and cost-effective imaging modalities. For example, a study by Grist (2019) found that lower field strength magnets may enable the development of economically feasible MRI systems and hence increase their value. The lower cost is due to a lower helium supply needed to cool a lower field magnet which can hence conserve helium supply and lower costs. In addition to lowering costs, they found that lower field magnet imagers can allow for the manufacturing of quieter imagers which can thus increase patient comfort levels. Further studies of this same kind have the power to boost patient outcomes and enhance cost efficacy through investigations of both past and recent MRI technological advancements.

CHAPTER SIX

What Science Is Involved In MRI Scans?

Melissa Speagle

An MRI machine is a large magnet which surrounds a patient's body and assists in the imaging of specific body parts. The adult human body is made up of about 60% water, a characteristic that is advantageous to this type of imaging (U.S. Geological Survey, n.d.). Water, also known as H_2O, is made up of two hydrogen atoms or protons, and one oxygen atom which are linked together by covalent bonds. The magnet in the MRI machine causes most of the protons, which are normally arranged in random orientations throughout the body, to align with a constant magnetic field (Jin, 1999). Radiofrequency (RF) pulses are then sent through the patient, causing the aligned protons to be pushed out of equilibrium and point at a 90 or 180° angle away from their aligned orientation. Turning off this pulse allows the protons to fall back into realignment, which represents a lower energy state (National Institute of Biomedical Imaging and Bioengineering, n.d.). To do so, these protons release electromagnetic energy that is detectable by sensors within the MRI machine. The amount of energy released by the protons and the time elapsed during realignment is dependent on the nature and surroundings of the molecules being analyzed. Radiologists can use this information to differentiate between types of tissues by comparing magnetic properties of the imaged regions. Contrast agents, typically containing Gadolinium, may be injected into the patient to quicken proton re-alignment. This allows radiologists to obtain a clearer image of the patient's body and is useful for examining vasculature and abnormalities, like inflammation, within the blood brain barrier (Preston, 2016).

NMR TECHNOLOGY

The complex science behind MRI technology can be fully understood only by physicists and mathematicians, but an attempt to provide a general overview will nevertheless be made. This story begins with the advent of nuclear magnetic resonance (NMR) technology, which was originally used for spectroscopic analysis of different chemical compounds (Jin, 1999). As mentioned in Chapter 2, the discovery of this phenomenon by Isidor Rabi paved the way for the development of the MRI scanner.

The nuclei of atoms placed in a magnetic field, which contain both protons and neutrons, can exist in an equilibrium between two energy states (Jin, 1999). The difference between a higher and lower energy state depends on the strength of the magnetic field. These nuclei can jump between the two energy states by absorbing or emitting a photon supplied by a light source. A radiofrequency probe creates the magnetic field that maintains these two states and subsequently, detects photons emitted by the atomic nuclei flipping between the two energy states. The energy difference between states can be altered by changing the frequency of the magnetic field in segments across the compound. This enables improved spatial resolution when using imaging technology.

A CLOSER LOOK AT PROTONS

The hydrogen atom is ideal for the application of NMR because it has only one proton (Jin, 1999). A single proton does not have a neighbour with an opposite spin to cancel its magnetic moment. This means that it must have a net spin and angular momentum. Furthermore, the ionic bonding between oxygen and hydrogen allows for hydrogen's nucleus to be deshielded from its single electron. This deshielding makes the nucleus more susceptible to a magnetic moment.

Protons have a dispersed positive charge that circulates about its axis as it spins, creating a current as the proton itself experiences angular momentum (Jin, 1999). This momentum results in the creation of a magnetic dipole moment: a small magnetic field generated by spinning net charge. The magnetic moment is expressed as a vector at an angle perpendicular to the current travelling around the proton's horizontal axis and parallel to the applied magnetic field (Nave, 2017a). Protons, similar to neutrons and electrons, each have an individual spin of ½ and a related magnetic moment, although this moment is smaller for protons than electrons (Nave, 2017b). The magnetic moment is measured in units called nuclear magnetons and requires the consideration of angular momentum. This value is equal to the z-component of the magnetic moment, which points in the same direction as the applied magnetic field. A gyromagnetic ratio may be used to describe the relationship between a magnetic moment and the angular momentum (Merriam-Webster, n.d.).

FELIX BLOCH ON NUCLEAR INDUCTION

In 1946, Felix Bloch published a ground-breaking paper on nuclear induction which would inspire the next generation of scientists to advance NMR research using his methodology (Bloch, 1946). Previously, magnetic resonance had been proven useful in describing the magnetic moments of neutrons and nuclei. It was known that the presence of an external magnetic field induces secondary spin on the nuclei, also known as a precession, which causes the nucleus to take a circular path around a constant magnetic field (Jones, 2020). This path might be compared to that taken by a spinning top. Larmor (precessional) frequency is the rate of precession of a proton's magnetic moment about an external, stagnant magnetic field. The frequency is calculated using the Larmor equation and depends on magnetic field strength. Radiation at the Larmor frequency applied to excited nuclei will allow these nuclei to realign in a parallel spin state, as this frequency represents the energy difference between parallel and antiparallel spin states (Jin, 1999). One former method to detect nuclear transitions included the deflection of molecular beams in a heterogeneous field (Bloch, 1946). Although not without its applications at the time, this method had a few issues that were addressed by Bloch's approach.

Bloch's experiments differed from previous ones in that he employed a large radiofrequency field to allow for a more significant alteration of nuclear moments and used a coil to measure the induced electromotive force (Bloch, 1946). The following will be a description of the conclusions drawn in Bloch's paper.

Nuclear induction was performed primarily using a weak RF field with numerous resonating Larmor periods; however, a less frequent, strong field pulse could be used as an alternative method (Bloch, 1946). Bloch's observations came from changes in macroscopic magnetic moments rather than those of individual nuclei, which frequently cancel out. Realignment of magnetic moments to reach thermal equilibrium was known to take between milliseconds and hours - a process termed relaxation time. Long relaxation times are not optimal, so paramagnetic catalysts can be used to quicken the process. Evaporation of oxygen and heating are two additional methods to shorten the relaxation time. Conversely, relaxation time can be extended through recondensation, catalyst removal, or cooling. Bloch suggested that the nuclear induction effect might be prolonged by first reaching thermal equilibrium in conditions that favour short relaxation time, followed by an extension of that time period. These results can be a source of comparison for magnetic moments and relaxation times of subsequent experiments. Bloch (1946) stated in his paper, "To study experimentally and theoretically this interesting relationship between nuclear relaxation time and atomic features seems to us, in fact, to be one of the fruitful fields of investigation which have now opened" (p. 461).

Several assumptions were made in conducting these nuclear induction experiments. Firstly, that any change of nuclear orientation was induced only by the external magnetic field; and secondly, that the external fields had a uniform influence on the sample (Bloch, 1946). This second criterion requires further assumptions that ignore the impact of atomic electrons (acceptable for water as the electronic spin moments are paired), intermolecular interactions between neighbouring nuclei, and thermal agitation. Although fields caused by thermal and intermolecular changes are much weaker than the external magnetic field, both can impact the magnetic moment, also known

46

as the polarization value. However, only thermal changes can affect the total energy of the spin system.

Bloch was the first to introduce the relaxation constants, T1 and T2 , and provided derivations for equations that connect these two values to the polarization value. T1 is the longitudinal relaxation time, also known as the thermal time constant (Bloch, 1946). This value can be easily swayed by thermal motion, which differs naturally across states of matter, and electronic structure. Thus, it is somewhat unpredictable. T2 is the transversal relaxation time constant and affects the other two planes of polarization that do not impact total spin energy. T2 does not have a substantial impact on T1 and can be impacted by internuclear forces, inconsistencies in the main magnetic field, or paramagnetic ions in solution.

Bloch (1946) concluded his paper by stating the advantages of nuclear induction over the use of molecular beams, such that it can be performed with little starting material and that it is useful in isotope analysis. He explained that nuclear susceptibility to magnetic moments and relaxation times can be studied separately in the future by selective variation of the resonant field or frequency with time.

MORE ON T1 AND T2

As mentioned earlier, T1 and T2 are parameters for relaxation times following an excitation pulse by a radiofrequency field. The decrease in the signal over time is denoted free induction decay (Constantinides, 2014).

T1 is a measurement of the time of longitudinal recovery following a repetition of pulses, which maximize magnetization of a material (Constantinides, 2014). In other words, it represents the speed at which a proton's spin returns to equilibrium, in parallel with the constant magnetic field, from an excited state. This value is important in differentiating between tissues in an MRI scan. An alteration in signal intensity could be caused by changes in T1, which could further indicate fluctuations in the local environment such as the surrounding macromolecules that are bound to water. For instance, lipids emit a higher signal than white and gray matter in the brain, which allows these two substances to be contrasted.

On the other hand, T2 is a measurement of spin-echo signal decay over time, which occurs due to local field inhomogeneities and resultant dephasing (Constantinides, 2014). This is the rate at which the protons lose phase coherence with the other nuclei that are spinning at a 90° angle to the constant magnetic field (Preston, 2016). It is important to note that an inconsistency in the external magnet will result in the quickening of free induction decay as protons will experience slightly different local magnetic fields (Constantinides, 2014). As the effects of such inhomogeneities are substantial and inevitable, T2* is the shorter time constant that is used more often to describe transversal relaxation (Jin, 1999).

T1 and T2 should be considered differently for solids, viscous fluids, and certain tissues compared to liquids (Constantinides, 2014). The molecules found in solids are comparatively frozen in place, so there is increased interaction between spinning protons and more rapid fluctuation in local magnetic fields. As a result, there is quicker loss of spin coherence in solids, and T2 in solids takes on a smaller value than T2 in liquids. Despite quicker loss of coherence in solids, there is a lower energy transfer between spinning protons and the solid lattice structure, so T1 is generally greater in solids than in liquids. For instance, cerebral spinal fluid has a lower T1 than body tissues with macromolecules, like complex sugars.

Besides differences in relaxation times between states of matter, some variation is dependent on molecule size (Constantinides, 2014). T1 is bigger in liquids with smaller molecules due to the more rapid movement and thus, poorer energy exchange between these molecules is present.

OBTAINING AN IMAGE

The localization of net magnetic moments, represented by vectors, is achieved through the application of mathematical equations to determine spin position (Constantinides, 2014). Magnetic field gradients of different strengths are used to excite each slice of the object being imaged. The resonant frequencies received by the MRI machine are related to the position of the nuclei or proton in space. Each gradient coil of the machine is independently regulated, with current being passed through it to form local magnetic fields. Different RF pulse sequences can be used to obtain the desired image (Preston, 2016). Repetition Time (TR) can be selectively varied, which is the time allowed between each RF pulse sent through a single slice of the object, as well as Time to Echo (TE), which describes the time from an RF pulse being sent out, to an echo signal being received. The FDA places limits on the strength of a magnetic field and the repetition times for clinical use or humans (Constantinides, 2014). Unfortunately, this places limits on resolution and speed of imaging. These gradient pulsations can create significant noise, largely due to Lorentzian forces on the conducting wires of the coils, which is the loud knocking noise that many patients hear when lying in the MRI machine.

Spin-warp imaging is the use of gradients that warp the precessional frequencies of protons (Constantinides, 2014). It is the most frequently used method of localizing spins for MRI imaging (Elster, 2021). One gradient is selected, typically along the z-axis, to excite a specific plane of the object (Constantinides, 2014). The other two planes orthogonal to the z-axis are obtained through information encoding. The x-axis is identified through frequency encoding gradients and the y-axis is usually obtained through phase encoding gradients. These transversal segments of magnetization become part of the data matrix known as k-space, which according to Moratal et al. (2008), "represents the spatial frequency information in two or three dimensions of an object." They explain that k-space data is related to the image data through Fourier transformation. The Fourier transform describes relative intensity levels which are then used to construct an image of the region of interest in a display of shaded pixels (Preston, 2016).

The relevance of T1 and T2 parameters is seen in that both values affect the magnetization vector, which the signal distribution used to construct the MRI image depends upon. (Constantinides, 2014). These values also affect image contrast. A scan is usually either T1-weighted or T2-weighted (Preston, 2016). A T1-weighted image is constructed with short TR and TE times, while a T2- weighted image applies long TR and TE times. T1- and T2-weighted scans are each mainly impacted by their corresponding T1 and T2 properties, respectively. For instance, for T1-weighted images, tissues with low T1 values appear bright and tissues with low T2 values appear dark (Constantinides, 2014). Radiologists may also use contrast to identify whether an image is T1-weighted or T2-weighted (Preston, 2016). One example is that cerebral spinal fluid appears darker for T1-weighted images and lighter for T2-weighted images.

Fluid Attenuated Inversion Recovery (Flair) is a third sequence used to produce MRI images (Preston, 2016). This technique is useful for detecting diseases through the appearance of abnormalities in the image. As with T2-weighted images, TE and TR are made to be quite lengthy, even to a greater extent, which enables these abnormalities to stand out brightly.

ADVANCED MRI TECHNIQUES

Variations of the basic MRI technique were inspired by the artifacts that are sometimes found in MRI images (Pagani et al., 2008). Image artifacts are observable features that are absent from the object being imaged and can be caused by mistakes in machine operation or natural body processes (Hornak, 2020). New techniques were developed by sensitizing MRI sequences to physical phenomena such as diffusion, blood flow, the magnetization transfer effect, and local field inhomogeneities (Pagani et al., 2008).

To begin, diffusion-weighted (DW) MRI is dependent on the diffusivity of water molecules through the medium or tissue being studied (Pagani et al., 2008). This property changes with the physiology and pathology of the region of interest and affects the signal received by the MRI machine. The signal decays as liquid molecules diffuse into heterogeneous, local magnetic fields. A diffusion coefficient can be calculated when various factors, such as tissue anisotropy, are considered. Additionally, diffusion tensor imaging can be used to describe diffusion tensor, the 3-D shape of diffusion (Huisman, 2010). This technique allows for the characterization of tissue anisotropy, direction of diffusion, and small structural or neuropathological changes in the brain (Alexander et al., 2007).

Another advanced MRI technique to note is called perfusion MRI, which measures cerebral blood flow (CBF) through tissue as a unit of time (Pagani et al., 2008). The injection of a contrast agent, such as Gadolinium, which measures the change in tracer concentration over time through the application of a $T2^*$ gradient. Alternatively, arterial spin labelling can be used in the absence of a contrast agent. This method inverts the longitudinal magnetization of an area upstream of the target. The movement of water molecules between blood and tissue causes relaxation of proton spins. CBF is determined by recording the ratio between longitudinal magnetization in the presence or absence of inversion and measuring the relaxation rate, which depends on blood flow.

Similar to the previously described techniques, magnetization transfer (MT) MRI relies on the magnetization of the protons found in water molecules. However, it differs in that it magnetizes them indirectly (Pagani et al., 2008). Firstly, surrounding macromolecules are magnetized using an RF pulse that excites spins at a frequency outside that which affects water. The affected spins are transferred to a group of water molecules bound to the macromolecules through chemical exchange or coupling of dipoles. The MT effect is expressed as a ratio of the percent variation of the signal between sequences with or without the saturation RF pulse.

Finally, functional MRI (fMRI) will be elaborated upon after its brief intro-duction in Chapter 3 and applications described in Chapter 5. To refresh, fMRI is a less-invasive alternative to PET scans, as it does not require a tracer (Pagani et al., 2008). Since brain activities require energy that is provided through glucose and oxygen, measuring the change in oxygen overtime can indicate regions of brain activity. Such information can be gathered in both rest and stimulating conditions. fMRI measures the change in ratio of oxyhe-moglobin and deoxyhemoglobin per unit volume of blood. Oxyhemoglobin is diamagnetic, while deoxyhemoglobin is paramagnetic, meaning it has an unpaired electron. This causes the magnetic resonance signal to decay more quickly as protons are more susceptible to dephasing. The BOLD effect is the blood oxygen level dependent contrast seen in the MRI image, which is pref-erably weighted for T2* decay.

CONCLUSION

Sometimes, it can be easy to get lost in the complex scientific details be-hind MRI and forget the simpler foundations. The large MRI machines that perform imaging depend upon the excitation and relaxation of microscopic protons! Credit is certainly due to the discoveries of magnetic resonance, NMR technology, and nuclear induction which laid the groundwork for the development of all the specialized MRI diagnostics available to us today.

CHAPTER SEVEN

What Have We Learned About MRI Scans In Recent Medical History?

Irene Fang

To this point, it has already been known that MRI scans are utilized to examine and study the structures of various organs and tissues of the human body through the generation of intricate, cross-sectional, 2-dimensional images, which are then combined to form 3-dimensional pictures. One may start to wonder, "What are MRI scans good at detecting? Are MRI scans accurate? Is it safe to undergo an MRI? What are some of the recent advancements of MRI technology?"

In recent decades, MRI has become a valuable and powerful imaging modality that is incorporated into clinical practice. They enable physicians to monitor the state of patients' health by identifying and distinguishing abnormalities revealed in the organs or tissues from their healthy states based on the size and distributions of bright and dark regions on MRI scans. Specifically, they are used to diagnose illnesses, injuries, and tumours, as well as to provide appropriate treatments based on these diagnoses. In particular, MRI is adept in detecting soft tissues, such as the brain; the metastases, or how far the cancer cells spread, to the brain; as well as the stage of some cancers that are otherwise hidden in CT scans, including uterine and prostate cancer (Stallard, 2019). A wide variety of organs and tissues, including blood vessels, nerves, bones, muscles, and ligaments can also be clearly visualized by the MRI, which other imaging techniques, such as CT scans and x-rays, were unable to detect (Stallard, 2019). Thus, MRI plays a crucial role in the diagnosis of shoulder and knee injuries (National Institute of Biomedical Imaging and Bioengineering, n.d.).

BENEFITS AND RISKS OF MRI SCANS

MRI techniques are advantageous because not only are they non-invasive, but no long-term health hazards are associated with its use (National Institute of Biomedical Imaging and Bioengineering, n.d.). For instance, patient exposure to deleterious ionizing radiation, which is used to create images in x-rays and computed tomography (CT) scans, is avoided in MRI scans (InsideRadiology, 2018). MRI scans will therefore be the preferred imaging approach for monitoring illnesses that require frequent imaging, especially brain diseases or injuries, for diagnostic and therapeutic purposes. Exposure to high radiation doses beyond certain thresholds can impair the functions of tissues and organs, thereby increasing the likelihood of the development of cancer (World Health Organization, 2016). According to the American College of Radiology, the maximum radiation one can be safely exposed to in a lifetime is 100 millisieverts, which is equivalent to 25 chest CT scans and ten thousand chest x-ray scans (Harvard Health Publishing, 2013).

However, despite being generally safe, MRI scanners are not suitable for everyone. For instance, it is unsafe for the patients with implants who carry metallic devices in the body, such as cardiac pacemakers, cochlear implants, insulin pumps, and artificial joints, to undergo MRI examinations since the strong magnetic field that is produced by MRI scans can cause the metals to dislocate in the body, rendering them malfunctional (U.S. Food and Drug Administration, 2017). Additionally, medical devices can negatively impact the quality of the MRI images, further degrading the diagnosis accuracy (U.S. Food and Drug Administration, 2017).

Before MRI scanning, gadolinium-based contrast agents (GBCA)—a generally safe liquid that enhances the clearness of MRI pictures and enables the radiologists to better visualize the organs or tissues of interest—is intravenously administered into an arm (InsideRadiology, 2018). The usage of GBCA in MRI scans can uncommonly induce transient allergic reactions, such as hives, nausea, and vomiting, for who are more sensitive to the contrast agents (InsideRadiology, 2018). Fortunately, the symptoms can be treated.
It is not optimal to administer GBCA for patients with moderate to severe renal impairments due to the decrease in their capabilities to excrete GBCA via urine from the kidneys (InsideRadiology, 2018). A potential nonallergic adverse effect of GBCA, such as the development of nephrogenic systemic fibrosis (NSF)—a rare, debilitating disorder that triggers skin hardening and an impairment of internal organs—in patients with severe renal failure undergoing dialysis was initially noted in 2006 (Marckmann et al., 2006). Noteworthy, merely associations, as opposed to definitive causal links, were established between NSF and GBCA retention in the tissues, indicating that the mere presence of gadolinium in the tissues does not conclusively cause NSF (Fraum et al., 2017). Some studies suggested that calcium ions in the plasma competitively bind to gadolinium-active sites, releasing the toxic free gadolinium ions that are not chelated into the plasma, thereby causing NSF; while others proposed that inflammation and infections were the fundamental risk factors that induce NSF (Budjan et al., 2014; Kuo, 2008). However, the exact underlying mechanisms of NSF remain uncertain today. Therefore, further controlled studies must be conducted to investigate the long-term effects of deposited GBCA to comprehend its health impacts on humans so that a maximum threshold for GBCA to be safely administered in MRI scans can be determined.

Another risk of MRI scans is that during the MRI scanning procedure, discomfort can be experienced by the patients suffering from claustrophobia as they are instructed to lie in a magnetic tunnel (National Institute of Biomedical Imaging and Bioengineering, n.d.). To help mitigate the distress, prescribed sedative medications may be taken prior to the procedure (National Institute of Biomedical Imaging and Bioengineering, n.d.).

In addition to the intimidating sensation caused by the enclosed space of conventional MRI scanners, loud noises that are 110 dB, or equivalent to the volume of a live rock concert, are generated by the machines (GE Health-care, 2019a). Essentially, the rapidly switched electric current flowing in the coiled wires produces magnetic fields, making the coils vibrate when images are being generated and obtained during the scanning procedure (Felmlee, 2005). The repetitive mechanical vibrations are not only audible, but are also further amplified by the hollow cylindrical bore (Felmlee, 2005). To mitigate this effect, earplugs are offered to the patients to diminish the risk of developing transient or permanent hearing loss, which is extremely uncommon. Reassuringly, the reduction of this acoustic noise is achieved in a late MRI technology—namely, Silent MRI Scan, which will be addressed in the "Recent Development in MRI technology" section of this chapter.

Radiofrequency (RF) radiation is a form of non-ionizing radiation that is emitted by MRI and is further absorbed by the body (American Cancer Society, n.d). Heat is generated through this process, which can lead to skin burns via immediate skin contact to the RF coils (Eising et al., 2010). Heating of conducting equipment, including pulse oximeter wires and some catheters, can also result in RF burns (Durbridge, 2011). The body, then, would adjust to maintain thermal homeostasis by dissipating the heat by means of convection and conduction (Durbridge, 2011). This is particularly difficult to accomplish by bariatric patients and newborns and thus, uttermost care must be provided by MRI technologists to minimize the hazard of burning (Durbridge, 2011).

Can pregnant women harmlessly undergo MRI scans? Is MRI scanning safe for the fetus? MRI scanning is mainly believed to be unharmful for the unborn in the second and third trimester of pregnancy (Bulas & Egloff, 2013). A cohort study conducted from 2003 to 2015 in Ontario aimed to address the safety of MRI scans and the contrasting agent to the fetus during the first trimester of pregnancy (Ray et al., 2016). This study encompassed 1.4 million childbirths and suggested that the risks of harm to the unborn in MRI-exposed women during the first trimester of pregnancy were not incremented compared to unexposed women (Ray et al., 2016). As well, although the presence of GBCA is not correlated with an elevated risk of congenital abnormalities in the newborn, rarely, MRI scans done with gadolinium are demonstrated to correlate with conditions resembling NFS, including infiltrative skin conditions in early childhood (Ray et al., 2016). Overall, though there are no known deleterious effects of GBCA to the fetus, extra precautions must still be taken, specifically in the first trimester, when organogenesis or the formation of organs of the fetus occur, which is an imperative stage for fetal development. Gadolinium can enter the fetal circulatory system and induce mutations, negatively impacting fetal health (Ray et al., 2016). Therefore, it is recommended to have gadolinium avoided during MRI scans for pregnant women, unless otherwise deemed absolutely necessary for the diagnoses and safe to both the patient and the fetus by the physicians (Ray et al., 2016). The risks and benefits must be discussed and clearly understood by the patients, and informed consent must be obtained from the patients to ensure patient autonomy is upheld.

ACCURACY OF MRI SCANS

With the progression of medical imaging technology comes increased intricacy and accuracy. Modern clinical applications of MRI scanning in the diagnoses of some diseases—endometriosis, autism spectrum disorder, and prostate cancer, to name a few, have been addressed in chapter 5 "Why are MRI Scans Worth Investigating?". The precision and efficiency in the diagnoses of cancer and multiple sclerosis will be further addressed in this section.

In current years, researchers have found that the MRI is a particularly useful screening instrument for the early detection of breast cancer in women who have higher susceptibilities of developing breast cancer due to genetic predisposition (Kriege et al., 2004). It is discovered that MRI scans have a higher sensitivity than mammography in tumour detection, meaning that MRI scans can more accurately identify a diseased condition in people who actually have the disease. Therefore, MRI scans provide a more holistic benefit in conjunction with the usage of other prevalent techniques for the screening of breast cancer, such as mammography and ultrasound (Kriege et al., 2004). Significantly, MRI scans can effectively facilitate early diagnoses and therefore increase the likelihood of successfully treating the disease and to enhance the quality of patients' lives.

In 2018, Filippi et al conducted a review summarizing the current findings of MRI application in patients with multiple sclerosis (MS), a chronic autoimmune disease that adversely impacts the brain and the spinal cord as one's immune cells begins attacking the myelin sheath, or fatty substance surrounding the neurons (McNamara, 2015). To diagnose MS, MRI scans have been utilized to supplement clinical assessments and to minimize the occurrence of misdiagnosis since the symptoms of MS can be similar to other diseases, such as stroke and migraine (Filippi et al., 2018). The combination of various MRI results, such as the ratio of image intensities in T1-weighted MRI and T2-weighted MRI, can be examined to elevate the sensitivity to cortical pathology or lesions constricted within the cortex (Righart et al., 2017). As such, although further studies must be conducted in determining the validity of MRI's specificity to myelin, currently, MRI has been a promising tool in providing reliable imaging measures that assist with MS diagnosis, monitoring how MS has progressed, and help in treatment development for the disease (Righart et al., 2017).

RECENT DEVELOPMENT IN MRI TECHNOLOGY

The power of MRI technology has come a long way over the past decade; in fact, it has revolutionized modern medicine in that MRI scans have become one of the most reliable tools in evaluating various disorders in humans from the pathological and anatomical details revealed by them (Harisinghani et al., 2019). MRI technology has been fueling multidisciplinary advancements in neuroscience, cardiology, oncology, and orthopedics. Computational progressions accompanied by hardware advancements have significantly improved the quality, spatial resolution, and contrast of the images, augmenting the usage of MRI in different organ systems (Harisinghani et al., 2019). For instance, an increase in the field strength, such as from 0.5-1.5T (teslas)—where MRI scans used to commonly operate a few years ago—to 3T, enables the background noise to be overcome by generating a stronger signal, hence producing clear images for smaller structures that are otherwise relatively difficult to detect, such as the nerve roots (Harisinghani et al., 2019; DMS Health, 2019).

To subdue one of the limitations of traditional closed bore MRI—patient discomfort experienced by those with claustrophobia—wide bore MRI scanners were introduced in 2004 (Imaging Technology News, 2019). In these new scanners, extra space is provided as observed in an increment in the size of the diameter of the bore, or the hole in the middle of the donut-shaped MRI scanner, by 10 cm relative to the conventional MRI (GE Healthcare, 2019a). Wide bore MRI scanners consist of wider or longer magnets surrounding the open space (GE Healthcare, 2019a). Several additional benefits provided by this new MRI technology include having an improved accessibility for a wider range of demographic which involves the obese population; having a shorter scan time and faster image processing compared to both closed bore MRI (open on only one end) and open bore MRI (open on both ends); and most importantly, high image quality is maintained due to a stronger magnetic field produced (Imaging Technology News, 2019). Having less time allocated to each examination not only increases patient throughput, but also contributes to an immense economic benefit (Imaging Technology News, 2019).

Most excitingly, a novel application of MR imaging: the first ultra-wide bore MRI scanner, also known as uMR OMEGA, was launched in 2019 (United Imaging, 2019). The uMR OMEGA consists of a 75 cm bore—that is, 5 cm wider than the wide bore MRI (United Imaging, 2019). Not only is the diameter larger, the table can now carry a greater weight, directly providing better care for pregnant women and those who require scans but were unable to fit into the bore of earlier models of MRI scanners by giving a more accurate diagnosis followed by a timely treatment (United Imaging, 2019). In addition, its more rapid scan time (within 5 minutes) renders uMR OMEGA an optimal MR imaging methodology for children, seniors, and claustrophobic patients, who cannot lay still for a long duration, since more ease is provided to them (United Imaging, 2019). As well, the usage of sedative medications to relax children and to reduce image distortion induced by motion artifacts during the MRI examination can be minimized, thereby preventing unwanted side effects of sedation, such as nausea. This fast-imaging feature also gives rise to innovative services as the usage of uMR OMEGA in an emergency setting becomes possible (Palmer, 2020). Pivotally, the image quality is not compromised by the larger diameter since 3T, a strong magnetic field strength, high homogeneity, meaning that the magnetic field is consistent, were utilized to generate clear images (Imaging Technology News, 2020; GE Healthcare, 2019c). Furthermore, injuries in the musculoskeletal system, including the joints, can be better examined in uMR OMEGA as provided by its high fat suppression property, effectively reducing the probability for image distortion (Imaging Technology News, 2020). Overall, by seeking means to enhance patient experience through alleviating anxiety and to overcome the obstacles that used to limit the access or reduce the willingness for some to complete the MRI examination, the medical field is striving towards its ultimate objective—that is, equitable healthcare for all.

With the aim to overcome another main drawback of MRI scans: the generation of intense acoustic noise, various approaches have been explored, including, but not limited to the installation of shielding materials surrounding the coils in the MRI scanners and the encompassment of the coils in a vacuum (Edelstein et al., 2005). In 2013, GE Healthcare pioneered the development of an innovative technology, also known as Silenz or Silent Scan technology, to dampen the noise to an ambient level that essentially goes unnoticeable compared with the conventional scans (Gintoft, 2013).

So what is the enigma behind Silent Scan technology? Simply put, sound attenuation is achieved by modifying how data is acquired—namely, by employing rapid RF switching in the coils as well as less abrupt amplitudinal changes in the gradient, while maintaining a fixed magnitude of the magnetic field levels (Matsuo-Hagiyama et al., 2017). Specifically, gradient refers to the magnetic field that is created by the electric current passing along the wires in an MRI scanner. However, due to the longer readout time for Silent MRI, critiques regarding image blurring are being raised (Holdsworth et al., 2018). To assess the image quality, Holdsworth and colleagues have found that the picture quality of the silent T1-weighted images were either comparable to or better than that of its traditional counterparts, permitting Silent Scan to be a reasonable alternative for the conventional MRI scanners. Correspondingly, another study has reported accomplishment in making minimal scanning noise with excellent image quality, including brain lesion diagnosis using T2-weighted MRI at 3T, reinforcing the capability for MRI scans to lower the noise without compromising the diagnosis accuracy (Ohlmann-Knafo et al., 2016).

This novel silencing technology effectively decreases the possibility of patient movement during the scan, ensuring a higher quality image to be obtained and decreasing the need to re-scan, further reducing healthcare costs. In essence, the scientific field is not only geared towards developing MRI technology that does not compromise image quality and diagnostic capacity, but is also progressively working towards establishing a more positive, supportive, safe, and patient-friendly healthcare system for all.

CONCLUSION

Despite being associated with several risks, engineers, hardware and software technicians, manufacturers, and healthcare providers have been industrious in creating MRI scans that minimize these limitations. Given the imperative role MR imaging plays in the diagnoses of various diseases in the 21st century, more research regarding the safety and efficacy of novel technological MRI advances must be conducted. Collectively, these professionals are turning the possibility of integrating multiple features: faster scanning; more efficient characterization of the diseased state in different organs and tissues; more patient-friendly and accepted by a broad-ranging demographic, into a single powerful MRI machine, into reality.

CHAPTER EIGHT

What Questions Are We Still Asking About MRI Scans?

Shea McMartin

MRI technology has made tremendous progress since it's first appearance in the medical scene in 1980. Since its discovery, there has been significant energy and research dedicated to the improvement of this technology. However, none of these advancements would be achievable without proper questioning from medical health professionals. With that in mind, this chapter focuses on diving deeper into the direction and growth of MRI treatments, questions that scientists and medical professionals have made in regards to the efficiency, costs, improvements of MRI coils, and how they are looking to further advance the software. Furthermore, this chapter will discuss the common questions from the perspective of the general public surrounding the accuracy, safety measures, comfort, wait times and alternative options when an MRI is no longer an option. That being said, before analyzing the deeper questions, there is a brief selection of the most common repeated questions sought after by the public.

COMMONLY ASKED QUESTION BEFORE AN MRI

There are several simple questions asked prior to a patient receiving an MRI that the majority are not aware of until actually enduring the process. The most commonly asked questions by the public are:

Will it hurt?
No, an MRI is a completely non invasive procedure, a patient will feel absolutely nothing during the process.

How long does it take?
The process of an MRI usually lasts approximately 45 minutes.

Why is an MRI important?
People commonly don't want to wait for the MRI appointment, however it is important for a doctor to properly diagnose a patient and determine if they have healthy or diseased tissue that cant be seen from the outside. This will in result improve the accuracy of the diagnosis and the establishment of a treatment plan (*MRI Frequently Asked Questions*, 2021).

PUBLIC CONCERNS AND DOCTORS APPROACHES TO RESOLVING THEM

Alongside the advancements of MRI technology, the test has become accurate at depicting the specific area of concern. However, the process of an MRI still creates difficulty for certain people if they cannot handle the noise, duration and being in an enclosed space for the duration. For the general public, many concerns arise in regard to their comfort and safety during the process. For young patients and infants in particular, it has proven to be difficult to lay still for the approximate 45 minutes it takes to do the scan, which is necessary for a proper imagine. In result, this can have further issues regarding the accuracy of the scan leading to less effective use for diagnosis.

NOISE

Doctors are looking to figure out new strategies regarding these obstacles. First of all, the noise, which is an acoustic noise created by rapid pulses of electricity, creating vibrations of gradient coils. This high intense noise is extremely loud and uncomfortable for the person receiving the MRI, as well as the staff providing the scan. For the majority this is bearable with the help of ear plugs, however the impacts of the noise may limit some patients abilities to continue with the scan. Due to this, doctors are trying to find solutions as to how they can minimize the noise made by the MRI, which will make the scan more available and comfortable to an increased number of people. A study was done by J.P. McNulty and S. McNulty to decipher the effectiveness of two different MRI systems and how their patients reacted/handled the noise. Participants that took place in the study were given a questionnaire for either system A or system B (the difference being the volume of noise with a variant of age and technology). The results came as followed: "The results of the questionnaire survey demonstrated significantly greater tolerance of the acoustic noise levels of System B (mean noise level rating of 2.45 on LIKERT scale) in comparison to System A (mean noise level rating of 3.71 on LIKERT scale) ($P = 0.001$). Significantly lower noise level descriptions were also demonstrated ($P = 0.01$). The maximum recorded sound levels also confirmed that System B was quieter than the System A" (McNulty & McNulty, 2009) (2).

Research on noise control is still being done, with a goal to figure out how to maximize effectiveness and clarity of the scan, while simultaneously creating an environment comfortable enough that the patient can sustain it.

SEDATION AND ALTERNATIVES

Furthermore, another significant concern is finding an effective way to have infants and small children during the process while sitting still. There are several different strategies currently being practiced in response to this consideration. For example, an article written by Andrea D. Edwards and Owen J. Arthurs explores "Paediatric MRI under sedation: is it necessary? What is the evidence for the alternatives?" (Edwards & Arthurs, 2011). (4) Sedation can be a controversial topic amongst certain families and beliefs, therefore some alternatives the researchers have found include neonatal comforting, sleep manipulation and adaptations to the physical environment. All of these options can never be considered 100% effective. However, sedation as strong and efficient it can be, there comes risks especially when dealing with the young patients.

"The main short-term risks of sedation / anaesthesia are those of under-sedation or over-sedation. Under-sedation is insufficient sedation to allow satisfactory imaging to take place, as judged by patient movement, repositioning or premature termination of the examination" (Edwards & Arthurs, 2011).

At an even higher risk is over-sedation, "respiratory depression requiring intervention, potentially resulting in the loss of protective reflexes and the ability to maintain an airway. Cardio-respiratory monitoring is mandatory in deep sedation / anaesthesia, but good clinical practice irrespective of conscious level" (Edwards & Arthurs, 2011).

Due to the effective results found with sedation it is continued to be practiced. However, the alternative approaches mentioned earlier create far less of a safety risk. Depending on the child, the parents and the paediatrician the route taken varies. Research continues to be done to find the most effective and safest path to MRI scans amongst young children/ infants (Edwards & Arthurs, 2011).

CLAUSTROPHOBIA

Another common issue comes from the enclosed space (creating feelings of claustrophobia), making it very difficult for any patience that struggles with claustrophobia to handle getting into the scanner let alone the duration of time. Research continues to try and discover a solution for this hurdle, to avoid potential anxiety attacks, trouble breathing, etc. that halt the process. However it is often strongly suggested to patients dealing with said fear to try and resolve it by meditation, medication, covering eyes, stretching and listening to music (Bergmann, 2021).

ALTERNATIVES

Furthermore, we know that MRI does not use radiation, granting a much safer environment for the general public. However, with the use of magnets during the process it is often questioned if pregnant women, or patients with any kind of metal implants are safe to continue with the procedure. Stanford Health Care (Risk Factors, 2021) (3) states the process cannot be done on patients with:

- *Implanted pacemakers*
- *Intracranial aneurysm clips*
- *Cochlear implants*
- *Certain prosthetic devices*
- *Implanted drug infusion pumps*

- *Neurostimulators*
- *Bone-growth stimulators*
- *Certain intrauterine contraceptive devices; or*
- *Any other type of iron-based metal implants*

Alternative methods of finding a diagnosis will have to be made. Furthermore they address pregnancy as a concern, "If you are pregnant or suspect that you may be pregnant, you should notify your physician. Due to the potential for a harmful increase in the temperature of the amniotic fluid, MRI is not advised for pregnant patients." (Risk Factors, 2021). Contrary to this article, thousands of women (*MRI Safety During Pregnancy*, 2019) have had MRIs performed who are pregnant and come out with no harm to themselves or the baby. It is commonly advised against however depending on the necessity the doctor will have to decide. In the case of a situation that completely deters the MRI from happening, there are alternative options that include ultrasound, X-ray scans, blood tests or a biopsy depending on the type of injury/symptoms being examined.

QUESTIONS BEING ASKED BY MEDICAL PROFESSIONALS TO MAKE FURTHER ADVANCEMENTS

Whereas safety and comfort is always a concern amongst the general population and medical professionals, doctors also take a harder look at improving efficiency of the procedure, the cost, wait times and how to create an even better more advanced image. The question that remains at hand is how to improve the technology already created, to make it faster, and more effective. There have been many recent advancements as stated in the previous chapter, however a major break though still being tested comes from MRI Coils array.

COILS

Gradient coils are used in MRI's to create vibrations in the main magnetic field (further explained in the latter chapter). The advancement of these coils could drastically improve MRI technology. These coils have been used since the MRI first was invented. However these are not the only coils that have been used. New, screen-printed, flexible MRI coils may be able to reduce the amount of time it takes to get an MRI scan. Researchers funded by the National Institute of Biomedical Imaging and Bioengineering (NIBIB), part of the National Institutes of Health (NIH), have developed light and flexible

MRI coils that produce high quality MRI images and in the future could lead to shorter MRI scan time periods." (Corea et al., 2016). Creating a shorter process would be beneficial for patients as well as doctors, by increasing accuracy of results due to the less time a person has being in a machine, as well as having greater the odds of positive results. However, with this new technology still being developed, it has been discovered that MRI coil arrays - do increase sensitivity, "but, current receiver coils are often anatomically unmatched for the human body; they are heavy, inflexible, restrictive, and are uncomfortable for many patients" (Corea et al., 2016). As the study continues it states Michael Lustig, Ph.D., and Ana Arias, Professors in the Department of Electrical Engineering and Computer Sciences at the University of California, Berkeley, have "developed new flexible MRI radiofrequency coils" that can be individually made to fit the size of the patient.

COST

Another constant question from doctors, hospitals and the general public have to do with figuring out the best quality treatment for the lowest cost. Technology such as the MRI machine can be very expensive, averaging a million dollars or more (*MRI Costs*, 2018). Due to this expense, it makes it difficult for hospitals to have access to as much supply as there is in demand. It is crucial the hospitals use of the machine is effective each time to stray from any waste. "according to results of a recent MRI global study of 40 radiologists and radiographers globally, 20 percent (or 1 in 5) of all MRI scans have to be carried out again because of patient motion, and this has a major impact on departmental efficiency.2 The three main clinical consequences of patient motion reported were:

- *74% decrease in image quality (cases of patient motion);*
- *70% possibility of exam not suitable for diagnostic purposes; and*
- *55% increase in the time required to carry out the scan" (Bringing More Value to Imaging Departments With MRI, 2019).*

These statistics are constantly being looked for ways to improve. As of 2020, MRI scans are costing "on average, an MRI can cost from $1,200 to $4,000" (2020) which is a very heavy cost for any facility, especially if needed to be carried out multiple times on one patient. Depending on one's personal insurance, they can also be a very heavy cost out of pocket for certain individuals. To continue, looking for solutions for effective use of funds, "according to recent research by the Advisory Board Company, one of the key imperatives for radiology leaders is to limit turnover and burnout of current workforce because of the tangible impact it has on quality, cost, productivity and health of radiologists.1 Instead of focusing solely on the technology itself, vendors now need to demonstrate how solutions provide improvement on the daily ways of working for the department" (Bringing More Value to Imaging Departments With MRI, 2019). The goal is to keep up with the volume of patients in need of an MRI versus the lack of access to get one. Another issue that continues to increase is the wait times to receive an MRI. Not only do these wait times hinder an individual from moving forward with an appropriate treatment but are increasing their risk of further injury. Unless considered an emergency, MRI wait times for Canadians often range from weeks to months. Having to wait months for access to a scan drastically can change the initial injury/diagnosis. Finding new strategies to lower cost, and efficiency of each individual scan will speed up the waiting process.

WAIT TIME PRIORITIZATION

Wait times for the scan are categorized in a five point classification system. These categories go from one being most important to four being least important, and the fifth being a specified procedure date. An Emergency scan is given to anyone who needs it to save a life or limb. Common MRIs are not used during life threatening cases due to their long duration time, therefore a CT scan is the preferred method due to its fast and efficient process. However in any case they can be performed they are prioritized. Priority number two is given to another emergency situation for a diagnosis but loss of life or limb is not a factor. Priority number three strains on the necessity for the scan to diagnose, but not urgent. It needs to be done sooner than stage four, but there are no expected severe consequences from a longer wait. Priority four is of least importance, still required for diagnosis, however this is where patients end up waiting weeks to months for their scan.

The system is similar to most other medical procedures, most urgent to least however the wait time is an increasing issue for more people are requiring scans than hospitals can promptly take (*National Maximum Wait Time Access Targets for Medical Imaging*, 2013).

CONCLUSION

In conclusion, the importance of constantly questioning new medical technology is crucial for the future advancement of medicine. MRI scans have greatly impacted the medical world and only continue to thrive with every advancement. There are many questions still to be answered, such as how to create a more comfortable environment will increase the accuracy of all scans. Finding solutions to every persons specific needs is an impossible task, however with every effort towards finding these solutions it makes it possible for more advancements to be made. The scientific research being done to improve the efficiency of a scan goes hand in hand with the efforts towards public need for comfort and safety. As these two elements progress, so will cost and quicker appointment dates that will inevitably result in helping more patients. Cost is constantly changing but remains a strong deterrent to the progression of MRI scans, however in the direction technology is heading the loss of funds over repeated tests for an inability of use for a diagnosis unclear imaging will minimize.

CHAPTER NINE

What Are The Parts Of An MRI Machine And How Do They Work?

Mohathir Sheikh

MRIs are complex machines and consist of several integral components required to create the magnetic resonance (MR) image. The main components being the primary magnet coil, shim coils, gradient coils, and radiofrequency coils (Durbridge, 2011). In this section, we will discuss the role of each component in the MRI machine and how they work together to create the MR image.

MAGNET COILS

The primary magnet in an MRI machine is the heart of the system and is also the largest and most expensive component. These magnets can weigh anywhere between 4500 kg to 7500 kgs and have a magnetic field strength in the range of 0.2 – 3.0 teslas (T) (Sammet, 2016). The role of the primary magnet is to generate a strong external magnetic field that forces the protons in the body to align with the field. Along with the strength of the magnetic field created, the precision of the magnet is also important. This is referred to as homogeneity which refers to the straightness in the magnetic lines within the iso-center (Durbridge, 2011). While inhomogeneities, or fluctuations, in the field strength can exist, they must be less than 3 parts per million (ppm). If the fluctuations are greater than 3 ppm within the scan region, distortions can be introduced into the image (Jin, 1999).

There are three types of magnets that have been used in whole-body MRI systems each with their own unique characteristics. The first type of magnet are resistive electromagnets, which were only used in the very early stage of MRI development and their design is now essentially obsolete (Kose, 2021). These magnets were a solenoid wound from copper wire. A solenoid is a coil of wire used to convert an electrical current into a magnetic field (Nicholson, 2018). The benefit of using resistive electromagnets is their low initial cost, however, they required considerable electrical energy to operate which resulted in significant costs over time (Murphy & Ballinger, n.d.). Additionally, these magnets require a cooling system and are far less powerful than the alternatives. With advancements in technology and design, resistive electromagnets are not commonly used in MRI machines today.

The second type of magnet used in MRI scanners are permanent magnets. These magnets are characterized as being composed of ferromagnetic metallic components such as iron or neodymium and can weigh over 30 tonnes (Murphy & Ballinger, n.d.). Almost all permanent magnet MRIs place the magnetic poles up and down and have a horizontal gap between them, creating a vertical static field (Kose, 2021). Typically, these magnets only produce a weak magnetic field between 0.2 to 0.4 T, which negatively affects their ability to be used in diagnostic imaging (Kose, 2021; Murphy & Ballinger, n.d.). However, their open design allows them to be more suitable for examining children and claustrophobic patients (Kose, 2021).

The third and final type of magnet found in MRI systems are superconducting magnets. Superconducting magnets like resistive electromagnets use a solenoid or cylindrical-shaped coil. The main difference in the superconducting magnets is the coils are made of niobium/titanium or niobium/tin alloys and surrounded by copper (Murphy & Ballinger, n.d.). A special property of these alloys is their zero resistance to electrical current when cooled down to about 10 kelvin. This unique advantage allows for large currents to be carried in very small superconducting wires without overheating (Sprawls, 2000). To create the superconductive magnet, a power source applies a gradually increasing current to the coils until the desired magnetic field is reached (Murphy & Ballinger, n.d.). Using liquid helium, the coils are then cooled to the superconducting temperature and then the power source is removed. The result is a closed loop of the electrical current in the coil without any dissipation of energy or heat and magnetic field that is always present (Sprawls, 2000). Since the liquid helium is exposed to extreme temperatures to keep the magnet cool, some liquid helium is boiled off into helium gas. The MRI coldhead, which is a mechanical device inside the MRI cooling system, recondenses helium gas back into liquid helium. Despite the closed loop, the liquid helium eventually needs to be replaced periodically as a very small percentage of it may be consumed through the daily use of the MRI machine. In the event the liquid helium levels drop too low or a ferromagnetic object enters the fringe field of the superconducting magnet, the current starts to reduce rapidly and starts producing heat. This event is known as a quench and results in the collapse of the magnetic field (Sprawls, 2000). In case of a quench of the magnet, the surrounding copper around the alloy coil which normally acts as an insulator at low temperatures will prevent the destruction of the coil. The major disadvantage of superconducting magnets are the high cost to replace liquid helium and the rigorous safety procedures to mitigate the risk of projectile events and quenches from occurring (Durbridge, 2011; Sprawls, 2000). The magnetic field created by the superconducting magnets is always on and it is not possible to turn them off like one could with resistive electromagnet magnets. Therefore, special care is necessary to prevent any unsecured ferromagnetic objects from entering the magnetic field and potentially becoming a dangerous projectile as they accelerate towards the magnet (Durbridge, 2011; Sammet, 2016). Despite these risks and disadvantages the superconducting magnets are the most common type of magnet used in MRIs due to the high magnetic field strength they can produce. Superconducting magnets at 1.5 T and above have the ability for MR spectroscopy, functional brain imaging and improved spatial resolution (Murphy & Ballinger, n.d.).

SHIM COILS

Optimal imaging requires a homogenous magnetic field across the image area and shimming is the process of adjusting the magnetic field to improve homogeneity. When magnets are manufactured and installed, some fluctuations may exist in the magnetic field. Therefore, metal shims may be installed to address these areas, and this is often referred to as passive shimming. Additionally, magnets also contain a set of shim coils. When patients enter the magnetic field, distortions can be produced in the form of inhomogeneities (Sprawls, 2000). This is due to the human body and other living tissue being magnetically susceptible materials (Durbridge, 2011). Adjusting the electrical current through the shim coils can help reduce inhomogeneities that can interfere with the imaging process, by generating a corrective magnetic field and is often referred to as active shimming (Sprawls, 2000). Active shimming is performed automatically through a computer on a patient-by-patient basis.

GRADIENT COILS

While the primary magnet creates a strong homogeneous magnetic field and the shimming coils work to reduce inhomogeneities already present in the magnet and produced by patients entering the field, the gradient coil's purpose is to deliberately create variations in the magnetic field. Gradient coils are thin loops of wire or conductive sheets found just inside the bore of an MRI machine. When a current is passed through the coils, a gradient field is created and distorts the main magnetic field in predictable spatially linear variations (Grover et al., 2015). There are three sets of gradient coils, which produce the distortions of the main magnetic field in the x-, y- or z-directions. These controlled variations in the magnetic field allow for localization of image slicing and frequency encoding which can be used to distinguish MR signals at different positions in space. The image slices can then be reconstructed into a three-dimensional MR image. Important elements of the gradient coils are their ability to be turned on and off quickly, and the strength of the magnetic field they can produce (Sprawls, 2000).

RADIOFREQUENCY COILS

The final component of the MRI system are radiofrequency (RF) coils. These coils are responsible for transmitting and receiving signals (Gruber et al., 2018). In general, these operations are split into separate coils, but in some cases, the two functions are combined into a single coil. While the combined coil has proven uses in applications with X-nuclei MR spectroscopy and ultrahigh-field MRI lacking a body transmitter, it lacks in optimization in coil design. Separating the transmitting and receiving functions into two coils allows the design of each coil to be individually augmented to optimize performance. Both the transmitter and receiver coils are resonance circuits that consist of a capacitor to store electrical energy and an inductor for the magnetic energy. Resonance circuits are formed when the capacitor and inductor are in parallel, which allows them to selectively respond to desired frequencies while discriminating against signals of different frequencies. In order for this to occur, the RF coils must be tuned and matched so the frequency of the electrical resonance of the coil circuit matches the frequency of the nuclear MR of the spins found in the tissue (Gruber et al., 2018).

As mentioned earlier the primary magnet produces the main magnetic field which is strong enough to align the spins of the protons along the direction of the field. The RF transmitter coil generates an RF pulse which produces a small magnetic field perpendicular to the main magnetic field. This excites the aligned spins off the main magnetic field and creates a signal that is picked up by the RF receiver coil. The more finely the coils are tuned, the stronger the signal picked up by the RF receiver. The analog sounds are then amplified using a preamplifier to a level that can be digitized. Unfortunately, sending a pulse through the RF transmitter once does not produce enough signal data to create an image (Sprawls, 2000). This is a process that must be repeated multiple times. Therefore, increasing the duration of the imaging process allows for more imaging cycles to be completed which provides additional signal data that will be used for the image construction. In general increasing the number of completed imaging cycles will directly improve the final image quality. Once all the signal data has been received, the computer system uses a mathematical process known as a Fourier transformation to create an image. This is a relatively fast process and doesn't have a large impact on the total imaging time.

SHIELDING

While this is not technically a component of the MRI machine, it is important to discuss the importance of using proper shielding in MRI environments. The problem with MRI systems is the large external magnetic field that exists around the primary magnet. This magnetic field can be extremely strong as mentioned previously and this can present two unique problems. The first problem is that the primary magnetic field can interfere with many different types of electronic equipment including computers and other imaging equipment located outside the MRI scanning room (Sprawls, 2000). The second problem is metal objects outside the MRI room, such as vehicles and building structures, can interfere with the magnetic field. These interactions can create distortions in the field and result in inhomogeneities being produced in the internal field. Therefore, it has become a common practice to install shielding around the MRI to help mitigate these problems. The general principle of shielding a magnetic field is to use conductive materials that can provide a more attractive return path for the external field as it passes from one end of the magnet to the other. There are two types of shielding: active and passive shielding. Active shielding is when additional coils are installed into the magnet assembly. When a current is passed through these coils, a magnetic field that directly opposes the external magnetic field is created, which reduces the size and strength of the external field. Passive shielding is surrounding the primary magnet with a conductive material to concentrate the external field through the shielding, effectively reducing the size of the external magnetic field.

Stray RF energy from the outside environment such as fluorescent lights, medical equipment, and radio equipment can also negatively impact the RF coil's ability to produce a high-quality image. These stray signals can be picked up by the RF receiver coil and cause distortions in the constructed MR image. Therefore, the room must be shielded against these possible interferences. Luckily, shielding against external RF signals can be done by simply surrounding the room in an electrically conductive enclosure. Interestingly, the thickness of the shielding does not impact the effectiveness of the shield. The most important factor is the entire room is enclosed by the shielding material with no gaps or holes. This requires special considerations in verifying the doors to the MRI scanning room contain the shielding material in order to create a closed environment during image acquisition.

SUMMARY

The primary magnet creates a strong magnetic field with the goal to align the spins of the protons in the sample with the main magnetic field. Within this magnetic field, there exists inhomogeneities that are present since the magnet was manufactured. Using shim sheets and pellets these areas of fluctuations are reduced through a process known as passive shimming. Active shimming refers to the use of shim coils to reduce inhomogeneities produced from the introduction of human tissue into the magnetic field. The shim coils are unable to completely eliminate all inhomogeneities, but they are able to reduce them enough from the scan region to avoid distortions in the resulting image. The gradient coils introduce inhomogeneities into the magnetic field in a predictable pattern. The resulting x-, y- and z-axis gradients allow for the creation of a three-dimensional image. Finally, the RF transmitter coil sends a pulse of energy in the form of a perpendicular magnetic field to the main magnetic field. This excites the aligned protons and causes the spins to briefly deviate from the main magnetic field. This signal is received by the RF receiver coil and amplified to a level that can be digitized, completing one image cycle. This imaging cycle is repeated multiple times in order to improve image quality. The attached computer system runs a Fourier transformation, which is a mathematical process, on the collected signal data to create the final image of the sample. Without proper shielding elements added to the MRI scanning room, the MRI system is subject to interference from external magnetic forces and stray RF signals. These external forces will create fluctuations in the external magnetic field and can produce inhomogeneities in the scanning region and can directly interfere with the RF coil's ability to receive the signals culminating in distortions in the produced MR image. Luckily, adding conductive material around the magnet assembly will help reduce the size of the external magnetic field. To further reduce the size and impact of the external field, additional coils can be installed into the magnet enclosure to create a field that directly opposes the main magnetic field. Finally, ensuring the room containing the MRI system is completely enclosed in a conductive enclosure, will block stray RF signals from interfering with the RF coils during image acquisition.

CHAPTER TEN

What Are Opposing Or Alternative Imaging Technologies To MRI Scans?

Anittha Mappanasingam

Medical imaging technology has improved drastically over the years. It is because of this improvement, that now when an individual goes into the hospital for a scan to determine the cause of an illness, or when a researcher is trying to find the source of a disease, they are able to discover the issue at greater ease than before. Currently there are many different imaging technologies that are readily available, and these options differ from one another in various ways, but mainly by their target, and the techniques used to operate the technology. This allows for healthcare professionals to combine more than one technique to accomplish their goal, whether that be for medical diagnosis or clinical research. Each of these techniques bring their individual characteristics to create an image that allows us to see what is going on inside the target region. MRI technology is one of the many imaging techniques that are available today but because MRI scans have specific features, they are not always the best option. This chapter will go through the other imaging technologies that are often used, such as X-rays, computed tomography (CT) scans, ultrasound, elastography, magnetoencephalography (MEG), radionuclide imaging techniques such as positron emission tomography (PET), and single photon emission computed tomography (SPECT), and optical imaging, and explore the similarities and differences between these technologies and MRI.

X-RAYS AND CT

X- rays were one of the earliest medical imaging techniques discovered and are still in wide usage today. X-rays are electromagnetic waves produced by moving electrons that are higher in energy, and smaller in wavelength than visible light (*X-rays*, n.d.). The colliding electrons that form x-rays interact with the atoms that make up our tissues, to form an image (*X-rays*, n.d.). These interactions however do not all occur with the same regularity. The frequency of the interactions are dependent on the atomic number of the matter that make up the tissue, and the amount of collisions (Kasban et al., 2015). It is the energy produced by these interactions that allow for either absorption or dispersion of x-rays (Chen et al., 2012). X-rays are absorbed by tissues when more collisions occur, and when the matter has a higher atomic number (Chen et al., 2012). A higher atomic number means more electrons, hence increasing the number of collisions. Similarly, x-rays are scattered by tissues when less collisions occur and when it has matter with a lower atomic number (Chen et al., 2012). Bones tend to absorb more x-rays than tissues, hence appearing highly contrasted. Meanwhile, less dense parts of the body, such as tissues, scatter X-rays, producing a lower contrast.

The MRI technique is more advanced than x-rays, and therefore there are significant differences between the two. X-rays are typically used as a structural imaging technique to detect bone fractures, metal implants, specific tumors, and abnormal masses (Chen et al., 2012). Meanwhile MRI imaging is branched out to serve the purpose of both structural imaging, through the basic MRI, and functional imaging, through fMRI. One of the biggest differences between x-rays and MRI lies in technique being used. X-rays use radiation to produce these images, while MRI uses magnetic fields (Kherlopian et al., 2008). In spite of using low doses of ionizing radiation, x-rays, when used multiple times, can pose side effects such skin reddening, and hair loss, hence limiting the repeated use of x-rays in humans (Kasban et al., 2015). Meanwhile MRI has no known negative side effects since it uses a magnetic field, as discussed in chapter seven (Kasban et al., 2015). Despite being slightly more advanced, and posing these differences, the two are predominantly similar. Ultimately, they both have similar overall goals, which is producing scans that allow individuals to see inside the body.

Another type of x-ray, CT, is a structural imaging technique that provides a 360 degree view of the body using computers and rotating x-ray machines (Kherlopian et al., 2008). CT scans provide more detailed images than the normal x-ray scans. Patients are placed in a hollow tube, where the x-ray portion of the CT transmits radiation around the patients (Kasban et al., 2015). These beams are then picked up by x-ray detectors (different from the x-ray tube), when they are leaving their body (Kasban et al., 2015). This information is then sent to the computer to produce the final product. Each time one complete rotation is made, a two-dimensional image is created (Kasban et al., 2015). Essentially, individuals are left with cross-sectional images that can then be put together to obtain a three-dimensional view of the target tissue (Kherlopian et al., 2008). This type of imaging is used to visualize bone, brain and tissues while normally being used to detect bleeding and tumors (Kherlopian et al., 2008).

In comparison to MRI, CT scans differ in terms of acquisition time, cost, quality of images, and health risk. When looking at acquisition time, which is the time taken to convert a signal into an image, CT scans tend to have lower acquisition times than MRI (Sivasubramanian et al., 2014). They are also relatively inexpensive in comparison to MRI (Kasban et al., 2015). Unfortunately, CT scans are not as detailed as MRI, increasing the likelihood of injuries going undetected (Purves et al., 2018, p.26). MRI provides an excellent soft tissue contrast which is a limiting factor of CT scans (Sivasubramanian et al., 2014). Lastly, similar to x-rays, CT scans also present the risk associated with high radiation exposure, limiting the recurrent use, while MRI does not pose this radiation risk (Kasban et al., 2015).

ULTRASOUND

Ultrasound is another fairly common medical imaging technology used in diagnostics and research. This imaging method involves acoustic coupling (Kherlopian et al., 2008). Essentially, it involves the use of an external hand held probe and a water-based gel to generate soundwaves that then produce an image on the computer (Kherlopian et al., 2008). The probe transmits sound waves into the body which are then reflected back and detected by the probe as an echo (Kherlopian et al., 2008). This echo is then interpreted by the computer to produce a two-dimensional image (Kherlopian et al., 2008). The computer converts these soundwaves into images by manipulating calculations that take the speed of sound and the time of each echo's return into account (Kherlopian et al., 2008). It is primarily used in cardiology, and helps assess blood flow in arteries and veins (Sivasubramanian et al., 2014).

When considering differences between ultrasounds and MRI, ultrasounds are only good at examining soft tissues, while MRI can be used for both soft and hard tissues (Sivasurbramanian et al., 2014). Ultrasounds provide real-time imaging, representing a significant advantage when time is restricted (Kherlopian et al., 2008). They are also relatively low-cost in comparison to MRI (Sivasubramanian et al., 2014). Unfortunately, unlike MRI imaging, ultrasounds cannot be used for whole body imaging (Sivasubramanian et al., 2014). On the other hand, similarly to MRI, ultrasounds do not expose individuals to radiation (Sivasubramanian et al., 2014).

ELASTOGRAPHY

Elastography is a medical imaging technique that recognizes tissues based on their elasticity (Kasban et al., 2015). Elastography is usually used as an addition to other imaging techniques such as ultrasound, MRI, optical imaging, and tactile imaging (Gennisson et al., 2013). In general, these elastography methods measure the stiffness of the tissues by monitoring changes throughout specific pathological processes (Sigrist et al., 2017). For example, an indicator of fibrosis is that the liver may become stiffer.

In ultrasound elastography, there are two main methods: quasi-static methods, and dynamic methods. Overall, quasi-static methods involve the application of a constant, normal external stress to the tissue (Gennisson et al., 2013). Calculations are then performed on the measurements derived from the normal strain creating a two-dimensional image (Sigrist et al., 2017). Dynamic methods implement time-dependent mechanical waves to the tissue (Gennisson et al., 2013). Ultrasound elastography is mainly used to detect diseases associated with the liver, kidney, thyroid, prostate, and lymph node (Sigrist et al., 2017).

Magnetic resonance elastography (MRE), as discussed in chapter three, involves placing pressure on the tissue with a device that projects vibrations onto the tissue at different rates depending on its elasticity, and is then captured by the MRI (Mariappan et al., 2010). MRE is typically used for examining changes in muscle associated with aging, and diagnosing breast cancer.

Regardless of the technique that elastography is paired with, it is noninvasive, quick, and extremely effective (Kasban et al., 2015). Hence, elastography is not necessarily more advantageous or disadvantageous than MRI, rather it provides individuals with a more refined assessment.

MEG

MEG is a functional imaging technique with its own unique distinctions. It involves detecting magnetic properties to observe electrical activity in the brain (Singh, 2014). Our brains create small, faint magnetic fields from electrical brain activity (Purves et al., 2018, p.28). However because these generated magnetic fields are so faint, MEG incorporates individual detector devices called superconducting quantum interference devices (SQUIDS) (Purves et al., 2018, p.28). SQUIDS require the maintenance of extremely cold temperatures, which is accomplished by liquid helium (Singh, 2014). SQUIDS are placed around a helmet which will be placed on the individual's head so that the detectors can pick up and amplify the magnetic signals (Purves et al., 2018, p.28). These magnetic signals are then used to create a three-dimensional map of the brain (Purves et al., 2018, p.28). MEG imaging takes place in a shielded room made of metal to block off external magnetic forces so it does not interfere with the faint magnetic signals being picked up by SQUIDS (Singh, 2014). This imaging technique is similar to, yet more developed than electroencephalography (EEG) which also measures brain activity (Singh, 2014). In the United States, most MEGs are used at epilepsy centres in epilepsy surgery for preoperative brain mapping (Singh, 2014).

MEG unlike MRI does not use radiation making it a better option for repeated use. MEG has a higher temporal resolution than fMRI, meaning that the results generated by MEG are more precise (Singh, 2014). However, MEG does not present a good structural view of the brain, and so this imaging method is often used in combination with MRI, which can provide structural imaging required for the clinical and research purposes (Purves et al., 2018, p.28). The combination of these two techniques is often referred to as magnetic source imaging (MSI) (Purves et al., 2018, p.28).

RADIONUCLIDE IMAGING TECHNIQUES: PET AND SPECT

Nuclear imaging techniques are imaging techniques that use radioactive tracers to view organ or tissue function (Kasban et al., 2015). The two main types of nuclear imaging techniques are PET and SPECT.

PET imaging uses radioactive tracers to show bodily functions in real time (Kherlopian et al., 2008). These tracers are created by injecting radioactive atoms into any biological compound used in the brain (Kherlopian et al., 2008). Once created, they are purified and subsequently injected into the bloodstream (Purves et al., 2018, p.27). Here, the radioactive tracers will release positrons, and the positron will undergo an annihilation reaction, which in this case, is the colliding of the positron with an electron (Kasban et al., 2015). The two photons, that are produced and released as a result of this reaction, are picked up by the PET scanner and provide individuals with the location of the tracer (Kasban et al., 2015). These events are put together to create a live view of brain activity. PET allows us to monitor detailed brain activity such as activity-dependent changes in blood flow, tissue metabolism, or biochemical activity (Purves et al., 2018, p.27). This form of imaging can be used to detect a wide variety of diseases because the tracers being used are dependent upon clinical application. For instance, when PET imaging is used to diagnose Parkinson's disease, fluorodopa is used as a tracer (Purves et al., 2018, p.27). Fluorodopa detects damaged dopaminergic neurons which is believed to play an important role in the development of Parkinson's disease (Purves et al., 2018, p.27). Although it serves as an exceptional diagnostic tool, it is rarely used in research or clinical applications due to the exposure of radiolabeled compounds (Purves et al., 2018, p.27).

SPECT is another type of nuclear imaging technique that involves the use of a radioactive tracer, and a CT (Purves et al., 2018, p.27). This technique allows for doctors to monitor the function of some internal organs. SPECT works very similarly to PET scans except they differ in the method in which radiotracers decay. Similar to PET imaging, when the radiotracer is administered to the patient, it remains attached to the biological compound until it reaches the target region where the radioisoptope tracer will begin to decay. As explained previously, while PET imaging uses positron decay, SPECT imaging uses electron capture decay (Kasban et al., 2015). This is when the nucleus is attacked by an electron in such a way that we lose proton, and gain a neutron (Gambhir et al., 2010, p.170). This process emits energy in the form of gamma rays, which is then detected by the SPECT scanner and produces the three-dimensional images of the body (Kasban et al., 2015).

MRI imaging is usually preferred over PET scans because they are less invasive and most cost-effective method for monitoring brain function (Purves et al., 2018, p.27). Although MRI has a better spatial resolution than these nuclear imaging techniques, SPECT and PET have a significantly higher sensitivity making it more precise to detect abnormalities in the body (Khalil et al., 2011). The information provided by nuclear imaging techniques are usually very accurate and specific due to this sensitivity (Kasban et al., 2015). In recent years, doctors have attempted to use nuclear imaging in combination with MRI to provide doctors with a better outlook on the bodily systems (Khalil et al., 2011).

OPTICAL IMAGING

A fairly unknown imaging technique is optical imaging. Generally, optical imaging is noninvasive and uses light to provide individuals with detailed images at the cellular level and tissue level (Kasban et al., 2015). There are various types of optical imaging such as endoscopy, optical coherence tomography (OCT), photoacoustic imaging, and diffuse optical tomography (DOT) (*Optical Imaging*, n.d.). As described in chapter one, endoscopy involves inserting a tube with a light source into the patient's mouth all the way to the digestive system, and is often used to evaluate diseases in the digestive system (*Optical Imaging*, n.d.). OCT works in a similar way to ultrasound imaging, except it uses light rather than sound to provide real-time images of tissue structure at a molecular level (Fujimoto et al., 2000). It is often used to diagnose or discover retinal diseases, and sometimes used to problems in cardiovascular diseases. Photoacoustic imaging, which is also similar to ultrasound imaging, involves laser-generated pulses to generate images based on the optical properties of tissues (Beard, 2011). This imaging technique is used to detect abnormal blood vessel growth, skin melanomas, and track blood oxygenation levels in tissues (*Optical Imaging*, n.d.). DOT involves the use of near-infrared light to observe the molecular properties of various tissues (Hoshi & Yamada, 2016). DOT is widely used in breast cancer imaging, and stroke detection (*Optical Imaging*, n.d.).

Optical imaging is an extremely effective detection technique for breast cancer and is often preferred over other techniques. These techniques also provide researchers and doctors with the ability to differentiate between soft tissues (Kasban et al., 2015). They are able to do this because different soft tissues have a unique way of absorbing and scattering light (*Optical Imaging*, n.d.). This allows for doctors to catch early indicators of abnormal functioning in the body. However, one major limitation is that these techniques tend to have a low spatial resolution, but to overcome this, it is often paired with MRI or CT imaging.

CONCLUSION

This chapter brought the various types of medical imaging technology together. MRI is an extremely necessary development, however, there are still limitations that MRI has that fortunately other techniques can solve. There are structural imaging techniques such as X-rays and CT imaging, and functional imaging techniques such as PET, SPECT, and MEG. Meanwhile, there are unique imaging techniques such as optical imaging that allows doctors and researchers to observe the tissues at a detailed, molecular level. Regardless, all these imaging techniques can be combined together to allow for an ideal view of the inner bodily system. The progression of medical imaging techniques is important in developing discoveries, and detecting diagnoses.

CHAPTER ELEVEN

What Misinformation Or Conspiracy Theories Exist Regarding MRI Scans?

Ashna Hudani

With the continual emergence of new and innovative medical technologies, it is important to ensure that accurate information regarding these technologies is accessible to the public, to prevent the spread of misinformation. There are several myths about MRIs which have been circulating publicly and provoking apprehension. This chapter discusses and debunks some of these myths using scientific literature, including the myths that MRIs cause exposure to radiation, and that MRIs are dangerous for people with tattoos.

MYTH 1: MRI EXPOSES PATIENTS TO HARMFUL RADIATION

Many medical imaging technologies, including x-rays, CT scans, and nuclear imaging, expose patients to ionizing radiation (Harvard Women's Health Watch, 2020). Ionizing radiation is high-energy wavelengths or particles that penetrate human tissue, which is used to create anatomical, physiological and functional images (Harvard Women's Health Watch, 2020; COCIR, n.d). This radiation can damage DNA and result in cancer-causing mutations (Harvard Women's Health Watch, 2020). MRIs, however, do not use ionizing radiation; instead, they rely on the body's natural magnetic properties to produce detailed medical images (Berger, 2002). Thus, because radiation is not used, there is no risk of radiation exposure during an MRI (Stanford Health Care, n.d).

MYTH 2: MRI CAN'T BE USED ON PATIENTS WHO HAVE JOINT REPLACEMENTS OR OTHER MEDICAL IMPLANTS

Joint replacement surgeries are common for patients suffering from conditions including injuries and osteoarthritis (GE Healthcare, 2019). The prosthesis, or artificial joint, can be made of metal, ceramic and/or plastic (GE Healthcare, 2019). Although many metals are considered MRI-safe, prosthetic implants, especially those containing metal, may distort MRI images by interfering with the magnetic field (GE Healthcare, 2019). Distorted images may include signal loss, failure of fat suppression, geometric distortion, and bright pile-up artifacts, which can ultimately lead to non-diagnostic examinations (Hargreaves et al., 2011; Sofka et al., 2003). However, new software options available on most commercial MRI machines can reduce the image distortion created by prostheses, by increasing receiver readout bandwidths, decreasing interecho spacing, reducing effective echo times, and using fast spin echo pulse sequencing (Sofka et al., 2003). These software modifications have been useful in reducing MRI image distortions in patients with orthopedic hardware, including hip, shoulder, and knee arthroplasties (White et al., 2000; Sperling et al., 2002; Sofka et al., 2003).

Similar safety concerns exist for patients with other implanted medical devices, including pacemakers and defibrillators. In the past, there have been patient fatalities when patients with older pacemaker models received MRI scans without appropriate programming or physician-supervised monitoring (Ferreira et al., 2014). However, a review of 15 studies, involving 1,419 MRI scans, reported no serious adverse effects for patients with pacemakers

(Zikria et al., 2011). Notably, 65% of these patients had MRI-conditional devices (Zikria et al., 2011). This classification was first introduced by the U.S. Food and Drug Administration (FDA) in 1997, after recognizing the need for standardized tests to address MR safety issues for patients with implants and medical devices (Shellock et al., 2009). Since then, the terminology used in the classification system has been revised such that implants and other devices are either labelled 'MR safe,' 'MR conditional' or 'MR unsafe' (Shellock et al., 2009). MR safe devices are those which pose no known hazards in all MR imaging environments, including nonconducting, nonmetallic and nonmagnetic devices (Shellock et al., 2009). MR conditional devices are those which have been demonstrated to pose no known harm to patients in specified MR environments with specified conditions of use (Shellock et al., 2009). All relevant parameters which may threaten the safety of the user must be included on the labels so that it can be taken into consideration (Shellock et al., 2009). Finally, MR unsafe devices are items that are known to pose hazards in all MR environments, including ferrous items (Shellock et al., 2009). These labels were taken into account in the Zikria et al. (2011) study, demonstrating that the risk to patients with pacemakers may be lower than imagined, if precautions are taken and specified conditions are being met (Ferreira et al., 2014).

MYTH 3: MRI CAN'T BE USED ON PATIENTS WITH TATTOOS

In the past, there have been reports that the strong magnetic fields used in MRIs can cause a tingling sensation or burns in patients with tattoos, by attracting the metallic pigments in tattoo ink (RAI, 2019). A recent scientific study at the University College London aimed to assess the risk of tattoo-related adverse events among patients undergoing an MRI scan (Callaghan et al., 2019). Between 2011 and 2017, a total of 330 participants between the ages of 18 and 66 were included in the study, in a total of 585 sessions (Callaghan et al., 2019). The inclusion criteria specified that participants should have at least one tattoo, their tattoo(s) should take up 5% or less of their body, their tattoos should be no more than 20cm in length, and that they should have no tattoos on their head, neck, or genitals (Callaghan et al., 2019). Of these participants, only one had experienced an adverse event, where a warm and tight feeling around their wrist tattoo during the MRI was reported, leading to termination of the MRI (Callaghan et al., 2019). This was classified by the researchers as a mild tattoo-related adverse reaction, which fully resolved within 24 hours without any medical intervention (Callaghan et al., 2019). Another participant reported feeling a tingling sensation following the MRI, but this was not classified as an adverse event, as the MRI was completed successfully (Callaghan et al., 2019). From this data, researchers estimated that the probability of a tattoo-related adverse event was between 0.17-0.30%, indicating a low risk under the specific study conditions (Callaghan et al., 2019). Although future studies exploring tattoo-related risks for larger tattoos, and tattoos on the head, neck and genitals may help to better inform the risk factors for patients with tattoos, this study demonstrates that patients with tattoos can safely undergo MRI scans.

MYTH 4: CONTRAST AGENTS ARE BAD FOR YOUR KIDNEYS

MRI contrast agents, administered either orally or intravenously, are used to increase the contrast difference between normal and abnormal tissues, improving the sensitivity and specificity of diagnostic images (Xiao et al., 2016; Ibrahim et al., 2020). However, MRI scans do not necessarily require contrast agents, and many continue to be performed without. New contrast agents are being discovered on an ongoing basis, and investigated for their applications, structures, mechanisms of action, pharmacokinetics, and pharmacodynamics (Xiao et al., 2016). Most MRI contrast agents currently used contain chelates of gadolinium, a rare earth metal (Ibrahim et al., 2020). Although atomic gadolinium is very toxic, gadolinium-based contrast agents (GBCA) have been used clinically since the 1980s and have a generally good safety profile (Ibrahim et al., 2020; Schieda et al., 2018). GBCAs used clinically should have a stable chemical structure to ensure that gadolinium is rapidly excreted from the body before it releases dissociated free gadolinium ions (Schieda et al., 2018). As a majority of the GBCAs are excreted almost exclusively by the kidneys, good renal function is imperative to patient safety prior to receiving a GBCA, and it has become standard practice for physicians to screen patients' kidney function prior to the administration of GBCAs (Schieda et al., 2018; Harvard Medical School, 2010).

Specifically feared among potential recipients of GBCAs is a condition known as nephrogenic systemic fibrosis (NFS), for which GBCAs have been identified as the causative agent, although there are some rare instances in which NFS was diagnosed in patients without known GBCA exposure (Schieda et al., 2018; ACR, 2021). NFS is a serious late adverse reaction associated with exposure to GBCAs, with symptoms including thickened and darkened areas of skin, shortening of muscles and tendons in the joints, and burning, itching and sharp pains in affected areas (Schieda et al., 2018; Mayo Clinic, 2019). At present, there is no consistently effective treatment for NFS patients (Schieda et al., 2018). Based on the evidence, NSF occurs almost exclusively in patients who have category G4 or G5 chronic kidney disease (CKD), because these patients have severely reduced renal function (Schieda et al., 2018). As such, because fetuses, neonates and infants have developing kidneys, these populations, as well as pregnant women, are considered at risk for NFS from the administration of GBCAs (Schieda et al., 2018). Another important factor to consider when discussing the risk of NFS is the specific type of GBCA being used (Schieda et al., 2018). Older linear GBCAs, including gadodiamide, gadopentetic acid, and gadoversetamide, have been associated with the greatest number of NFS cases (Schieda et al., 2018). These agents are considered to be high risk for causing NFS in at-risk populations, which include patients on dialysis, patients with severe or end-stage CKD, and patients with acute kidney injury (AKI) (Schieda et al., 2018; ACR, 2021). The use of these three agents on the aforementioned classification of patients remains contraindicated by the FDA, European Medicines Agency, and Health Canada (Schieda et al., 2018). Newer linear agents of GBCAs, including gadobenate dimeglumine and gadoxetate disodium, however, are associated with zero unconfounded cases of NFS (Schieda et al., 2018). Although the specific differences in risk for each GBCA requires more investigation, these findings suggest that because of the different properties of each GBCA, some agents will be more probable to result in a higher risk for patients with compromised kidney function (Schieda et al., 2018). In any case, it is a myth that contrasting agents are bad for your kidneys. As explained above, patients with predisposed risk, including those with impaired renal function, may develop NFS from GBCAs. However, for patients with normal kidney function, who can rapidly release the contrast agent following the scan, NFS is not likely to occur.

MYTH 5: MRI CAN'T BE USED TO DIAGNOSE PATIENTS WITH CLAUSTROPHOBIA AND ANXIETY

Claustrophobia, which is the combined feeling of suffocation and restriction, can be a major deterrent for patients who require an MRI scan (Iwan et al., 2020). This can lead to negative health outcomes, including decreased chances of early diagnosis and treatment for patients with certain diseases which are detected by MRIs (Iwan et al., 2020). It is estimated that approximately two million scans around the world cannot be performed, either due to premature termination or patient refusal due to claustrophobia (Munn et al., 2015; Enders et al., 2011). A systematic review and meta-analysis showed that 1.2 people out of 100 have a claustrophobic event during an MRI scan (Munn et al., 2015). Completing MRI scans on these patients may require that they are sedated, which can be life threatening (Enders et al., 2011).

Most MRI machines are imagined to be what older generation models resemble - closed, narrow, and long bores (Iwan et al., 2020). However, MRI models vary in design, including machines that are open panoramic scanners, short bore scanners, open one-column scanners, and upright open scanners (Iwan et al., 2020). Less restrictive design of scanners is one approach to reduce claustrophobia in patients receiving MRIs (Munn & Jordan, 2013; Harris et al., 2004; Tillier et al., 1997; Quirk et al., 1989). The initial visual impression of an MRI scanner is a relevant contributing factor to feelings of claustrophobia, suggesting that new scanners with visual scanner design features can influence cognitive attitudes about MRIs (Iwan et al., 2020). In a recent study, 160 patients at high risk for claustrophobia were surveyed to investigate which MRI scanner designs patients prefer (Iwan et al., 2020). The most common areas for MRI design improvement found in this study were noise reduction, more space overhead, and more space overall. These patients visually preferred open-bore MRI designs over shore-bore MRI designs. Notably, another study, which included a large sample size of over 55,000 patients, concluded that newer short-bore MRI machines with 97% acoustic noise reduction may reduce the incidence of claustrophobia in patients by a factor of three (Dewey et al., 2007). These results, although varied, demonstrate that MRI design can reduce claustrophobia experienced by patients, and better accommodate their needs.

Beyond MRI design, there is extensive literature suggesting that claustrophobic and anxious patients may experience better outcomes with the implementation of patient-centred radiology practice, which emphasize interpersonal interactions (Ajam et al., 2020). Because of the pressure that many radiology departments face, in meeting productivity targets and reducing costs, less attention is given to patient interactions and maximizing patients' comfort (Ajam et al., 2020). One strategy to improve patient outcomes is through effective communication with patients prior to the MRI scan (Ajam et al., 2020). This claim is supported by a survey-based study, which found that patients who received information about the MRI and what to expect prior to the scan had significantly less anxiety scores than those who had not received the same information (Munn et al., 2015). In another study, with a sample size of 33, a group of patients receiving an MRI who received a set of verbal statements prior to the scan and received communication through a built-in intercom every two minutes during their scan had lower anxiety than the control group, as measured through serum cortisol changes (Tazegul et al., 2015). In addition to providing information, one study found that providing relaxation techniques with information was more effective in reducing anxiety than just providing information alone (Quirk et al., 1989). Among relaxation techniques, music has been found to reduce MRI-related anxiety during scans (Földes et al., 2017; Walworth, 2010). Additionally, cognitive techniques including guided imagery and sounds can reduce patients' experiences of anxiety during an MRI scan (Lukins et al., 1997). These strategies can be implemented without significant costs and resources, and improve the experiences for anxious and claustrophobic patients receiving MRI scans, leading to better patient outcomes.

CONCLUSION

This chapter overviewed and debunked five pertinent myths regarding MRI scans including: MRIs expose patients to harmful radiation; MRIs can't be used on patients with joint replacements or other medical implants; MRIs can't be used on patients with tattoos; contrasting agents are damaging to patients' kidneys; and MRIs can't be used to diagnose patients with anxiety and claustrophobia. MRI technology is anticipated to advance in future years, leading to increased developments in their use. At the same time, researchers will continue to investigate these topics, which will lead to new developments in knowledge about MRIs and their applications. To avoid consuming and reproducing inaccurate ideas which can have harmful health implications, it is important for the general public to consult reliable and accurate sources of information and to consult their physicians for specific contextual information.

CHAPTER TWELVE

How Are MRI Scans Talked About Commonly And In Popular Culture?

Sifar Halani

It's interesting to note that the medical imaging technique used in MRI scans is known as nuclear medical resonance imaging, or NMRI, for short (Frank, 2016). However, in almost every use case, the nuclear aspect is dropped from the title, leaving just medical resonance imaging, or MRI. While nothing in the process changes from the addition or subtraction of a letter in its title, the impact on patients and their friends/families is positive overall. Many people are afraid of nuclear technology, and would not be open to such technology being used in their bodies (Frank, 2016). This fear is reasonable, as nuclear technologies have been responsible for massive socio-political, economic, and health-related catastrophes. These range from nuclear power plant failures and nuclear missile crises to the effects of intense radiation on human bodies and the environment.

WHY ARE THEY CALLED MRI?

The Chernobyl disaster of 1986 is one such example. Over 30 people died to severe radiation exposure within 2 months after the Chernobyl nuclear reactor failed, 350,000 people had to be permanently relocated from their homes, and to the present day, researchers have tracked over 5,000 cases of cancer resulting from the increased radiation levels emitted into the environment from the reactor (World Nuclear Association, 2020). Additional research shows that children who were exposed to the radiation from Chernobyl are still at higher risk of cancer and tumour growth (Cardis & Hatch, 2011). This disaster was deemed a one-off, as the safety protocols and staff training were neglected due to Cold War pressures, and the plant's design itself was one-of-a-kind (World Nuclear Association, 2020). However, another major nuclear disaster occurred in Fukushima, Japan in 2011, with over 160,000 people forced to permanently relocate. While no people died from the initial reactor failure, it is estimated that 10,000 people have either already died, or are still in the process of dying from exposure to radiation (Aliyu et al., 2015).

These two disasters, in addition to the greater conflict caused by the Cold War, including the Cuban Missile Crisis, have given a negative connotation to nuclear technology in everyday life, and as such, it is reasonable that the medical community removed the term from MRIs. In this case, it is better to appease people and help the public feel more comfortable with helpful medical imaging scans than to be precise in terminology. Part of this fear comes from the fact that knowledge of these disasters and conflicts is widespread, and well-known, whereas the foundations and principles of nuclear technology are less so. From electromagnetic theory and quantum physics all the way to biological diagnostics, the intricate knowledge networks needed to piece together fundamental information in order to create an MRI machine are so deep, that very few people could adequately trace and follow the path from foundational physics all the way to a complete MRI scan (Frank, 2016).

In North America, over 40 million MRI scans are performed each year (Advisory, 2019). This means that general knowledge and awareness regarding MRIs is relatively high in comparison to the overall population. When someone says that they are going to get an MRI, people will tend to know what they are talking about. Either through passing conversation, on TV or in movies, or through personal experience, most people have at least heard the term MRI being mentioned. This, however, does not mean that most people understand how an MRI machine works, what its purpose is, and the related benefits and risks. This combination of knowing the reality of an MRI but being unaware, or unconvinced, of all the facts leads to increased excitement and fear, which in turn leads to some interesting developments in the field of radiology and the pop culture conversation surrounding this medical imaging technique (Health Imaging, 2020).

THE CONTROVERSY SURROUNDING GADOLINIUM

One aspect of MRIs that has been present in pop culture recently is the use of contrast agents, namely gadolinium, in improving the visibility and effectiveness of the scans. The administering of metallic elements into the human bloodstream is relatively common, with platinum used in cancer therapeutics, technetium in cardiac scanning, and gadolinium in MRI scans (Caravan et al., 1999). When needed, a small amount of gadolinium is inserted into the bloodstream, allowing the MRI machine to detect molecule movements easier and produce a more useful image. Every year, over 3 metric tons of gadolinium is used for MRIs, with each patient receiving about a gram per scan. Additionally, over 30% of MRI scans use contrast agents, and the percentage is expected to rise further (Caravan et al., 1999). In the decades since its inception, gadolinium contrasting has been used to help diagnose over 300 million patients worldwide (Patterson, 2017).

However, this does not mean that the use of gadolinium is free of controversy. In 2017, actor Chuck Norris filed a lawsuit in California on behalf of his wife, Gena Norris. They claimed that Gena suffered lethargy, burning, and pain due to the gadolinium that was administered to her during a scan (Patterson, 2017). They also filed multiple other lawsuits on behalf of others who claimed they were also suffering following the use of gadolinium in their scans. Since blood and urine testing for the contrast agent were not available until recently, Norris and his team claim that gadolinium manufacturers failed to inform customers about the risks associated with its use (Patterson, 2017). The symptoms that Gena Norris identified have recently been given a new title, Gadolinium Deposition Disease, or GDD for short. This disease was originally described as Nephrogenic Systemic Fibrosis, or NSF for short, but was then split by a group of nephrologists (Semelka et al., 2016). In 2016, the non-fatal and treatable version of the disease was given its more modern name of GDD, and new symptoms were included in the definition, such as muscle wasting, burning pain, shaking, confusion, and kidney damage (CBS, 2017). The original, incurable and potentially life-threatening version retained its name of NSF. Following scientific and operational improvements such as more stable gadolinium-based compounds and refraining from using gadolinium on patients with advanced renal failure, no new cases of GDD were reported after 2009 (Semelka et al., 2016).

The lack of new cases led people to believe that the gadolinium-based compounds used in MRI scans were extremely safe and did not lead to GDD. It also led people to believe that patients with normally-functioning renal systems would not develop the disease (Semelka et al., 2016). However, it has since been seen that patients with normal renal functions still reported severe symptoms of GDD shortly after receiving gadolinium insertions and even beyond the expected time frame for symptom appearance (Burke et al., 2016). Since the rise of GDD cases, multiple online support groups have formed for people suffering from gadolinium-derived toxicity symptoms, not limited to GDD (Semelka et al., 2016). These support groups allow people suffering from gadolinium-related symptoms to come together and share their burdens and hopes, while meeting others who are going through similar situations (MRI Gadolinium Toxicity, 2012). Additionally, some of these groups conduct advocacy efforts to raise awareness for people who suffer from GDD, and petition to Federal and State governments and organizations in the USA to provide more support for those with GDD and other gadolinium-related symptoms, and to provide more information and warning ahead of time for patients undergoing contrast-agent assisted MRIs (Williams & Grimm, 2014). Since the advent of these support and advocacy groups, numerous publications have been made regarding GDD's prevalence and simultaneously, its relative obscurity. Notably, representatives of the GDD support community have presented in front of the FDA, leading to an FDA investigation on the effects of gadolinium-based compounds on the brain, and the greater effects of gadolinium retention on the human body (Williams & Grimm, 2014). While the FDA released its first safety announcement regarding gadolinium in 2015, health care professionals are still allowed to use gadolinium contrast imaging, but have been cautioned to use it only in circumstances that deem it necessary (Williams & Grimm, 2014).

The obscurity of GDD, NSF, and other gadolinium-related diseases and toxicity symptoms makes it difficult for those suffering to get the proper treatment and care they need. Chuck and Gena Norris' attorney mentioned that it is difficult to get a proper diagnosis, and that the Norris', and others they are representing in lawsuits, have been misdiagnosed with Lyme disease, ALS, and other diseases (CBS, 2017). Once the other diseases had been ruled out, only then was the verdict of GDD reached. There is a difficult situation at hand, where patients are worried about developing complications and diseases on the one hand, and also need the contrast-aided MRI scans on the other hand in order to help diagnose and treat other diseases and issues that they currently face. Gena Norris spoke about how she sees MRIs as critical to patient care and diagnosing, and does not want people to stop choosing MRIs as viable diagnostic tools. However, she stated that she wished she was warned about the use of gadolinium, and would have chosen to undertake the MRI without the use of contrasts had she been given the full information (CBS, 2017). As of now, the FDA has not found conclusive proof of harm caused by gadolinium remaining in the body after the scan, but has taken the step of changing drug labels to add additional warnings and improved processes for certain high-risk patient groups (CBS, 2017).

From the perspective of radiologists, nurses, technicians, and other healthcare workers, gadolinium, affectionately referred to as "gado," is still trusted as one of the safest drugs ever developed (Caravan, 1999). Sometimes also known as "magnetic light", gadolinium is readily available in hospitals and clinics. It is no more difficult to "ask for some gado" than it is to start a saline drip or obtain a blood sample (Caravan, 1999). As MRIs become increasingly common, more specialists are becoming familiar with "gado", and now neurologists, cardiologists, urologists, ophthalmologists, and other types of physicians are experimenting with gadolinium to find new ways to improve their diagnosing (Caravan, 1999). While there are contrast agents for MRIs based on both iron and manganese, gadolinium remains the most popular.

This popularity is reflected in the fact that more MRIs are performed in North America than in other parts of the world (Advisory, 2019). While certainly a mark of technological achievement and proficiency, an abundance of MRI imaging is not necessarily a good thing. For instance, excessive imaging can result in unnecessary information being given to healthcare professionals, leading to increased worrying. This stems from the risk that MRIs will uncover "radiographic incidentalomas", which are abnormal situations discovered through imaging which did not serve as the original reason for undergoing the imaging in the first place (Advisory, 2019). Finding such incidentalomas may not yield any measurable medical benefits, but notifying patients of them may lead to increased financial expenditure and mental stress. This is because when the patients are informed of such incidentalomas, they themselves may begin to pursue additional treatment and diagnosing. Additionally, the previously mentioned risks associated with contrast imaging agents, namely gadolinium, also arise when patients opt to undergo additional voluntary imaging (Advisory, 2019). Research is now leaning towards introducing the concept of "diagnostic waste" in medical education. This concept provides healthcare workers with a framework for determining whether a certain test or scan is necessary, what the effects and consequences on patients may be, whether there are other options available, and what information of value the test will uncover (Advisory, 2019).

CINEMAVISION GOGGLES

While the medical ethics and healthcare risks of MRI scans and their associated contrasting agents can be controversial in everyday conversation, the prevalence of MRI technology and usage has led to numerous interesting developments. One of these developments is the CinemaVision goggle. These goggles were developed to help people who experienced fear and anxiety during MRIs by letting them watch a movie of their choice during their scan. The goggles themselves look similar to virtual-reality glasses and provide an immersive experience, helping drown out the loud noise of the MRI machine. This is especially helpful with children, and prevents the need to use sedatives (Cosgrove, 2013). A major cause of the fear and anxiety that patients experience during MRIs is the loud noise that the machine makes when it is working. This noise is, in practice, often at a level of 125 decibels, or in MRI terms, three teslas. This equates to being at a rock concert, or having a balloon pop right beside your ear (Ray, 2012). This noise is caused by the vibration of metal coils in the MRI scanner reacting to strong pulses of electricity. The immersive nature of the CinemaVision goggles helps reduce claustrophobia, and often, the movie is started before the patient is even placed in the MRI machine. Anxious and claustrophobic reactions from patients can disrupt the scan, slowing down the process and making the test inaccurate. This can also lead to the need to conduct additional re-scans. By helping the patients become less aware of the MRI taking place, healthcare professionals can conduct faster and more accurate scans, which then leads to better diagnoses (Cosgrove, 2013).

While through CinemaVision goggles, movies can be used to help patients make it through MRI scans and increase the scans' effectiveness, movies can also lead to fascinating discoveries through the use of MRIs as well. A research team interested in neural activity used CinemaVision goggles inside MRI scanners to determine the brain's reaction to different types of fear, the parts of the brain most active during this fear, and what kind of fear caused the most activity (O'Conner, 2020). The MRIs showed the researchers that there are two types of fears experienced by the brain, a sustained, anxious, and uneasy fear, and an acute, sudden shock fear (Hudson et al., 2020). Their MRI scans determined that sustained and acute fear follow separate, distinct neural pathways, and that psychological horror films based on true events, and containing unknown factors produced the most fear in the brain (Hudson et al., 2020). In the future, horror filmmakers could potentially test scenes and plots of their films using MRIs to help them position their movie with the desired level of scariness and type of fear produced.

NEUROMARKETING

This is an application of the cutting-edge field of neuromarketing. Through neuromarketing, companies and individuals can test upcoming products on sample groups through imaging their brain activity in response to given stimuli. The stimuli could be advertisements, soft drink flavours, songs, movies, toys, or any other conceivable product. The marketers could then study the results and improve their products accordingly (Ariely & Berns, 2010). This type of marketing through MRIs has the potential to be more cost-effective and lead to deeper insights than what is possible through more traditional marketing methods such as surveys and feedback forms. Neuromarketing researchers have already experimented with tests using MRI machines to help compare people's neural responses to Coca-Cola vs Pepsi in a number of different controlled scenarios, such as being blind tests, labelled, and mis-labelled (Castro, 2011). Additionally, MRIs have helped music researchers and marketers identify key features in songs that trigger brain responses leading to people liking the song and having it be catchy enough to be "stuck in your head". This knowledge was used to predict which pop songs would be hits in the future, and found that many hit pop songs create the same sort of brain response, and those that don't are less likely to achieve the same level of popularity (Castro, 2011).

MRI IN MOVIES AND RESEARCH

Beyond being a venue for movie consumption, MRIs have also recently been prominently featured in movies themselves. In 2018's popular action-thriller film "Venom", Tom Hardy's starring character undergoes an intense scene in an MRI machine (Gay, 2018). In the story, the character has been infected with an alien parasite called Venom, who lives within his body, giving him superhuman abilities but causing him to act in unnatural ways. The character's girlfriend brings him to a radiologist who proceeds to do an MRI scan. In the film, the machine's noise is dramatically amplified in order to add creative flair and build suspense. Hardy's character is severely anxious during the scan, and it is soon revealed why. The Venom creature has an adverse response to the MRI scan and actually removes itself from his body inside the machine, making for one of the most intense scenes in the entire movie (Gay, 2018). While on the one hand, this type of portrayal does increase fear and anxiety surrounding MRI scans, it should also be noted that the MRI did succeed in its task. It identified the creature living inside the man and even removed it from his body, leaving the man extremely fatigued but alive. In this sense, it is both a cinematically appropriate portrayal of an MRI for audience engagement, and still an endorsement for the effectiveness of the procedure (Gay, 2018).

The public's fascination at the ability to scan your brain and body, and pop culture's willingness to embrace the MRI and turn it into a public spectacle has brought many creative research projects to life. A hearing specialist placed jazz musicians and freestyle rappers in MRI machines and had them perform, in order to determine how they are able to improvise their music for extended periods of time (Fung, 2012). He determined that they are able to shut off certain parts of their brain, allowing them to experience less self-monitoring. A woman gave birth inside an MRI machine, allowing doctors to make precise scans of the changes in both the woman's and the baby's body during the birth process. An actor was given a piece to read normally inside an MRI machine, and then given a piece to read in-character, allowing researchers to identify what changes in an actor's brain when they are performing (Fung, 2012). People have played video games inside MRI machines in order to help determine how the brain reacts to virtual puzzles and challenges. People have agreed to enter MRI machines and then have snakes or tarantulas inserted into the machine with them to help measure the brain's reaction to creatures quite commonly considered scary to many people. People have had sex inside MRI machines to help scientists study brain activity during intercourse and orgasm (Fung, 2012).

CONCLUSION

The public's general comfort level with MRIs tends to vary widely depending on personal experiences, beliefs, and factors. This is not without cause. There are a large number of people who claim to suffer severe symptoms from MRI contrast agent exposure, and increased warnings and safety protocols are being implemented. In one case, an MRI scan caused a gun to malfunction and fire by itself inside the machine (Beitia et al., 2002). MRI machines cause fear and anxiety through the loud noises they make, and most medical workers refrain from mentioning the nuclear aspect in order to reduce fear. On the other hand, MRI machines have helped millions of people around the world obtain accurate diagnoses and have undoubtedly helped save many lives. They have also paved the way for new industries such as neuromarketing, and have allowed researchers to be creative in ways that would otherwise be impossible. For good or for bad, the MRI has been ingrained in our popular culture.

CHAPTER THIRTEEN

Where Is MRI Research Headed In The Future?

Rayna Kamal Naik

The 20th century has brought forth many significant advancements in medicine and medical technology alike. MRI is often cited as one of the most important developments among these advancements as its impact in biomedical research and clinical diagnosis has continuously grown into the 21st century (Grist, 2019). Although it is considerably new technology in the grand scheme of things, over the last 10 years, MRI technology has established itself as one of the preferred diagnostic imaging tools compared to its counterparts (Sobol, 2021). With that said, it should be noted that the use of MRI greatly depends on the particular medical condition that is under investigation (Sobol, 2021). In order to provide a sense of just how far this technology has come, the findings from MRI research studies have contributed to better image quality, faster exam times, improved throughput, increased patient comfort, and have also introduced novel scanning techniques that continue to change the standards of diagnostic imaging practices today (Sobol, 2021). The final chapter of this book will take a detailed look at new developments within MRI research and examine what we can expect from ongoing and forthcoming studies in the near future. Topics of discussion include ultrashort echo-time sequence, MAGnetic resonance image compilation (MAGiC), GOBrain, and ViosWorks (Roach et al., 2016; GE Healthcare, 2017; Kadom & Trofimova, 2019; Zagoudis, 2018).

ULTRASHORT ECHO-TIME SEQUENCE

MRI has long been established as the diagnostic imaging method of choice for the investigation of various systems, including the nervous system and the musculoskeletal system (Roach et al., 2016). In spite of this, completion of respiratory system MRIs have proven to be a difficult task due to the low proton density of the lungs and sensitivity to physiological motion (Roach et al., 2016). To fully grasp this concern, it is important to consider the lungs from a molecular point of view. MRI scans depend on hydrogen atoms in order to produce accurate images. Hydrogen atoms exist in copious amounts within areas of the body that are filled with water and fat (Berger, 2002), namely, the brain and muscles. However, the production of images through MRI for the respiratory system has been a challenge since air flows through the lungs, resulting in a lower density of hydrogen atoms. Since these hydrogen atoms are required to produce MRI scans, this presents a significant barrier in the production of successful images of the lung. This is simply one of multiple factors that have resulted in decreased image quality and barriers in accurately identifying structures around the lung (Roach et al., 2016).

Furthermore, there exists the issue of physiological motion sensitivity. This refers to sources of noise that compromise the image quality and may interfere with the interpretation of the scan (Havsteen et al., 2017). High image resolution is typically associated with high sensitivity to motion artifacts (Havsteen et al., 2017); thus, it is essential for any individual undergoing an MRI scan to ensure minimal motion artifacts in order to produce an adequate image. A motion artifact simply refers to a "patient-based artifact that occurs with voluntary or involuntary patient movement during image acquisition" (Murphy, 2020). This phenomenon has the potential to increase the total scan time, which may cause a cyclical pattern whereby motion artifacts are further aggravated (Havsteen et al., 2017). Physiological noise falls under the category of involuntary patient movement, and may include sources such as cerebrospinal fluid, flux in subarachnoid space surrounding the spinal cord, cord motion itself (Fratini et al., 2014), swallowing reflex, small spontaneous head movements and cardiac pulsation (Havsteen et al., 2017). Related to the respiratory system, breathing motion is one of the primary sources of artifacts in MRI acquisitions that may negatively impact diagnosis (Honal et al., 2010). For these reasons, CT scans have typically been the go-to method for lung imaging (Dezube, 2019).

Evidently, there have been prominent challenges in transforming MRI as the standard for lung diagnostic imaging. Regardless of this, important developments have certainly been made to optimize lung MRIs—this includes the use of ultrashort echo-time (UTE) MRI technology. According to a press release, UTE allows imaging specialists to take accurate images of tissues that typically disappear quickly in areas like the lungs (Canon Medical Systems, 2015). A study completed by Roach et al. (2016) compared UTE methods with CT methods as a biomarker of lung structure abnormalities in children with early stages of cystic fibrosis (CF). The rationale for this study was that recent advancements in MRI technology have improved our ability to assess lung function in patients with CF and may function as a tool that can be used to enable positive patient outcomes (Roach et al., 2016). The researchers concluded that UTE MRI was successful in identifying structural lung disease in young patients with CF (Roach et al., 2016), and the results provided similar contrasts to the images produced by CT (Kumar et al., 2016).

UTE provides a specific advantage in lung imaging: it enables high quality three-dimensional scans of the whole lung in an accelerated manner (Miller et al., 2014). This ultimately allows for image acquisition at very short relaxation times (Kumar et al., 2016). In patients with CF, UTE has been particularly successful in portraying the structural lung features required to evaluate disease, including: bronchial wall thickening, bronchial dilation, mucus plugs, emphysema, cysts, abscesses and atelectasis (Dournes et al., 2020). This can be compared with conventional MR imaging, which was not as successful as UTE in depicting these features (Dournes et al., 2020).

MAGNETIC RESONANCE IMAGE COMPILATION (MAGIC)

One of the setbacks of conventional MRI scanning is the time it takes to produce a satisfactory image, as it typically takes around half an hour to complete (GE Healthcare, 2017). While many individuals are able to take the time spent in the bore to relax, others may feel claustrophobic, ultimately decreasing their comfort level and preventing the completion of scanning. MAGnetic Resonance image Compilation (MAGiC), a multi-contrast MR technique, is the product of a collaborative effort between GE Healthcare and SyntheticMR whereby specialized software develops eight contrasts from a single scan at a faster rate than conventional scanning techniques (Imaging Technology News, 2016). This provides healthcare professionals with flexibility by allowing them to adjust the images retrospectively, thus saving more time and lowering costs as the need for rescanning decreases (SyntheticMR, 2017). When using traditional imaging methods, a rescan may be needed if appropriate contrast was not acquired initially (Imaging Technology News, 2016). This would require the patient to undergo further scanning, thus utilizing more time and money. The use of MAGiC will enable clinicians to produce several image contrasts in a single scan, including T1-, T2-, STIR, T1 FLAIR, T2 FLAIR, Dual IR, Phase Sensitive IR, and Proton Density weighted images in a single session (Imaging Technology News, 2016). Since this new technology is in the form of software as opposed to machinery, it is compatible with existing GE MRI machines, eliminating the need to purchase new ones (GE Healthcare, 2017). Approximately 30 percent of MRIs are completed for the purpose of neurological evaluation, i.e. for the brain. This is the primary reason that MAGiC was created for brain use (GE Healthcare, 2017).

A blinded, multi-center, multi-reader clinical study was conducted in the United States that compared MAGiC to traditional MR imaging (Imaging Technology News, 2016). To briefly summarize the materials and methods section of the study, a mixture of MAGiC images and conventional images were collected from 6 sites and 109 patients (Imaging Technology News, 2016) to be analyzed by 7 blinded neuroradiologists (Tanenbaum et al., 2017). Conventional images included l T1- and T2-weighted, T1 and T2 FLAIR, and STIR or proton density scans, while MAGiC images consisted of synthetic reconstructions from multi-dynamic multiple-echo images (Tanenbaum et al., 2017). The images were then evaluated for diagnostic quality, morphologic legibility, radiologic findings and artifacts (Tanenbaum et al., 2017). The results of the study indicated that overall diagnostic quality of MAGiC images were "noninferior" to traditional imaging (Tanenbaum et al., 2017). Noninferior in the context of this research study means that both imaging techniques were found to be comparable in terms of diagnostic quality.

You may be wondering, if they are comparable in diagnostic quality, what was the point of creating MAGiC in the first place? To answer this question, diagnostic quality refers to the extent to which clinicians are able to reliably assess the scans. The results of this study indicated that both MAGiC and traditional imaging methods were comparable, so clinicians were able to assess the scans in an analogous manner. With that said, MAGiC technology was also able to reduce overall scan times, provide the reader with novel information (Tanenbaum et al., 2017) and increase cost savings. Therefore, the findings of this research indeed hold clinical value.

GOBRAIN

To understand the need for GOBrain, it is important to describe some facts and figures that provide sufficient background. According to a study by Kadom & Trofimova (2019), the number of MRI studies which require sedation on behalf of children increases by 8.5% annually. Although this is not inherently negative and the sedation of children for imaging studies has been demonstrated to be safe, the concern lies in the potential neurotoxicity of specific anesthetics (Kadom & Trofimova, 2019). In fact, until recently, scientists did not understand the specific mechanisms that underlie general anesthetics (Okinawa Institute of Science and Technology, 2020). In addition to neurotoxicity, sedation adds both monetary and time costs (Kadom & Trofimova, 2019). Due to the increased amount of time spent at the imaging center, children may experience motor imbalance, gastrointestinal issues, agitation and restlessness (Kadom & Trofimova, 2019). Thus, fast MRI sequences help individuals avoid the need for sedation, and are especially significant for use with younger populations (Miller & Smith, 2017). GOBrain is a fast MRI protocol which enables a push-button brain examination within 5 minutes of initiation (Raplino & Heberlein, 2016).

Moreover, a paper published by Miller & Smith (2017) discussed the use of GOBrain, a fast-brain MRI technique, for children with shunt-dependent hydrocephalus. This is a condition that develops shortly after the onset of subarachnoid hemorrhage (Yim et al., 2018). Hydrocephalus is characterized by an abnormal accumulation of fluid in the ventricles that exist within the brain, causing the ventricles to expand, thus increasing pressure on brain tissue (National Institute of Neurological Disorders and Stroke, 2020). Children with this condition who opted into imaging studies were often required to undergo frequent scanning, which incorporated both CT and MR imaging. The downside to this is that CT scans involve exposure to radiation, which may exhibit potentially harmful effects, while brain MRIs are very time consuming and demonstrate increased sensitivity to motion artifacts alongside the need for possible sedation (Miller & Smith, 2017). More recently, however, fast MRI sequences have been more accessible for a greater number of conditions such as macrocephaly, intracranial cysts, congenital and non-congenital anomalies, and postoperative follow-up (Miller & Smith, 2017).

While a number of fast-protocols have been used in the past, they typically use a single type of pulse sequence that carries potential drawbacks, such as the inability for these pulse sequences to characterize tissue and fluid (Miller & Smith, 2017). This highlighted the need for enhanced image quality using fast-MR protocols, which can be completed through the use of multiple pulse sequences and planes (Miller & Smith, 2017). The research by Miller & Smith (2017) incorporated the MRI protocol with the addition of the GOBrain 5-minute sequence towards the end of the procedure. The study concluded that fast-imaging techniques in a pediatric context were able to decrease scan times and motion artifacts, as well as enable the patient to steer clear of sedation. Fast imaging methods were also found to be noninferior to the conventional protocol in terms of diagnostic quality (Miller & Smith, 2017).

Other benefits were demonstrated through a study conducted by Kadom & Trofimova (2019), such as increased patient throughput, as the 5-minute brain MRI protocol opened up time on the MRI schedule for alternative use. Although the image quality of the 5-minute protocol was reasonable, it may not meet standards that pertain to certain imaging findings such as depicting small parenchymal lesions or small blood products (Kadom & Trofimova, 2019). This is due to the image acquisition constraints of achieving fast scan times. For this reason, Kadom & Trofimova (2019) recommend restricting the use of this MRI protocol to patients who report symptoms of a "headache" or "migraine" as the independent sign, with a low risk of exhibiting structural brain pathology.

VIOSWORKS

ViosWorks was developed by Dr. Albert Hsiao in the basement of Stanford Hospital in response to the limited options for cardiac MRIs (GE Healthcare, 2019). It is an MR software that is able to complete a full scan of cardiovascular anatomy, function and flow in approximately 10 minutes or less; this is in contrast to conventional cardiac MRI exams that typically take over an hour (GE Healthcare, 2019). This software is able to effectively conduct a three-dimensional chest volume scan while adding four more dimensions — myocardium motion, blood flow, time and fully automated quantification

(Zagoudis, 2018). As a result, there are a total of seven dimensions: three in space, three in velocity, and one in time (GE Healthcare, 2019). Each scan successfully retrieves around 20 gigabytes of information in a free-breathing exam (Zagoudis, 2018). The scans reveal blood flow in the heart in the form of a moving graphic, and enables users to move the image for a view at an alternative angle (GE Healthcare, 2019). This helps doctors determine whether blood is pumping normally through the heart or whether there are any irregularities (GE Healthcare, 2019). ViosWorks' software helps speed up the cardiac imaging process while producing high-quality images. Patients are no longer required to hold their breath during the process, and the rapid speed of exams enable hospitals to accomodate more patients at once (GE Healthcare, 2019).

CONCLUSION

This final chapter explored various advancements and enhancements of existing MR imaging software and infrastructure. Evidently, there have been large strides in the development of new techniques that have streamlined the imaging process for both clinicians and researchers. The overall advantages of these methods include: decreasing the amount of time needed to produce clear scans, increasing patient throughput, and optimizing image quality. Specifically, we focused on ultrashort echo-time sequence, MAGnetic resonance image compilation, GOBrain and ViosWorks, each of which resolve specific issues in MR imaging. Ultrashort echo-time sequence has enabled the rapid production of high quality three-dimensional scans of the lung, while MAGnetic resonance image compilation has provided clinicians with the opportunity to adjust brain scans retrospectively through the production of eight contrasts from a single session. Furthermore, GOBrain has increased the speed of scans which reduces the need for sedation in populations that are unable to undergo a sitting with minimal movement, such as pediatric populations. Finally, ViosWorks has paved the way for faster, more detailed cardiac MRI scans in seven dimensions which also allow patients to freely breathe during the scanning process. Considering that MRI technology has only existed for approximately 40 years, we can be certain that these techniques will account for a small fraction of overall developments, and have laid the foundation for new and innovative advancements in radiology that are yet to follow.

Conclusion

MRI scans have been, and will continue to be, an essential part of the diagnosis process for a vast variety of conditions. Its development has allowed clinicians to identify and monitor the progression of various conditions at greater ease than before, and exposed patients to less risks compared to previous diagnostic imaging techniques. As a relatively new invention, it has faced its fair share of skepticism over the years, but with its widespread adoption, many of these myths and fears have been quelled. With new and exciting innovations emerging all the time, the applications and benefits this technique provides is only bound to grow in the future.

References

ACR. (n.d.). ACR Manual on Contrast Media (ISBN: 978-1-55903-012-0). Retrieved 2021, from https://www.acr.org/-/media/ACR/files/clinical-resources/contrast_media.pdf.

Advisory Board. (2019, January 23). America performs far more CT and MRI scans than other countries. Advisory Board. https://www.advisory.com/daily-briefing/2019/01/23/scanning.

Agosta, F., Valsasina, P., Rocca, M. A., Caputo, D., Sala, S., Judica, E., & Filippi, M. (2008). Evidence for enhanced functional activity of cervical cord in relapsing multiple sclerosis. Magnetic Resonance in Medicine, 59(5), 1035–1042. https://doi.org/10.1002/mrm.21595

Ahmed, H. U., Kirkham, A., Arya, M., Illing, R., Freeman, A., Allen, C., & Emberton, M. (2009). Is it time to consider a role for MRI before prostate biopsy?. Nature Reviews Clinical Oncology, 6(4), 197–206. https://doi.org/10.1038/nrclinonc.2009.18

Ajam, A. A., Tahir, S., Makary, M. S., Longworth, S., Lang, E. V., Krishna, N. G., Mayr, N. A., & Nguyen, X. V. (2020). Communication and Team Interactions to Improve Patient Experiences, Quality of Care, and Throughput in MRI. Topic in magnetic resonance imaging : TMRI, 29(3), 131–134. https://doi.org/10.1097/RMR.0000000000000242

Alexander, A. L., Lee, J. E., Lazar, M., & Field, A. S. (2007). Diffusion tensor imaging of the brain [Abstract]. Neurotherapeutics, 4(3), 316–329. https://doi.org/10.1016/j.nurt.2007.05.011

Aliyu, A. S., Evangeliou, N., Mousseau, T. A., Wu, J., & Ramli, A. T. (2015, September 29). An overview of current knowledge concerning the health and environmental consequences of the Fukushima Daiichi Nuclear Power Plant (FDNPP) accident. Environment International. https://www.sciencedirect.com/science/article/pii/S016041201530060X-?via%3Dihub.

American Cancer Society. (2019, May 16). MRI for Cancer. American Cancer Society. https://www.cancer.org/treatment/understanding-your-di-

agnosis/tests/mri-for-cancer.html#:~:text=MRI%20creates%20pic-
tures%20of%20soft,is%20or%20isn't%20cancer.

American Cancer Society. (2020, May 5). Tests for Brain and Spinal Cord
Tumors in Adults. American Cancer Society. https://www.cancer.org/
cancer/brain-spinal-cord-tumors-adults/detection-diagnosis-stag-
ing/how-diagnosed.html.

American Cancer Society. (n.d.). Radiofrequency (RF) Radiation. https://
www.cancer.org/cancer/cancer-causes/radiation-exposure/radiofre-
quency-radiation.html#:~:text=What%20is%20radiofrequency%20
(RF)%20radiation,remove%20electrons%20from%20an%20atom.

American Chemical Society. (2011). NMR and MRI: Applications in Chemistry
and Medicine. Retrieved from American Chemical Society: https://
www.acs.org/content/acs/en/education/whatischemistry/land-
marks/mri.html

Anagnostou, E., & Taylor, M. J. (2011). Review of neuroimaging in autism
spectrum disorders: what have we learned and where we go from
here. Molecular Autism, 2(1), 1–9. https://doi.org/10.1186/2040-
2392-2-4

Anzai, Y., Minoshima, S., & Lee, V. S. (2019). Enhancing value of MRI: a call
for action. Journal of Magnetic Resonance Imaging, 49(7), e40–e48.
https://doi.org/10.1002/jmri.26239

Ariely, D., & Berns, G. S. (2010, March 3). Neuromarketing: the hope and hype
of neuroimaging in business. Nature. https://www.nature.com/arti-
cles/nrn2795.

Baltzer, P. A. T., Kapetas, P., Sodano, C., Dietzel, M., Pinker, K., Helbich, T. H., &
Clauser, P. (2019). Contrast agent-free breast MRI: Advantages and
potential disadvantages. Der Radiologe, 59(6), 510–516. https://doi.
org/10.1007/s00117-019-0524-7

Bartlett, D. J., Burkett, B. J., Burnett, T. L., Sheedy, S. P., Fletcher, J. G., & VanBu-
ren, W. M. (2019). Comparison of routine pelvic US and MR imaging
in patients with pathologically confirmed endometriosis. Abdominal
Radiology, 1–10.

Beard, P. (2011). Biomedical photoacoustic imaging. Interface Focus, 1(4), 602–631. https://doi.org/10.1098/rsfs.2011.0028

Beitia, A. O., Meyers, S. P., Kanal, E., & Bartell, W. (2002, May). Spontaneous Discharge of a Firearm in an MR Imaging Environment. American Journal of Roentgenology. https://www.ajronline.org/doi/full/10.2214/ajr.178.5.1781092.

Bell, D. J. (2020). Paul Lauterbur. Retrieved from Radiopaedia : https://radiopaedia.org/articles/paul-lauterbur

Bell, D., & Murphy , A. et al. (2020). Peter Mansfield. Retrieved from Radiopaedia: https://radiopaedia.org/articles/peter-mansfield?lang=us

Berger, A. (2002). Magnetic resonance imaging. BMJ (Clinical research ed.), 324(7328), 35. https://doi.org/10.1136/bmj.324.7328.35

Berger, A. (2002). Magnetic resonance imaging. BMJ (Clinical research ed.), 324(7328), 35. National Center for Biotechnology Information. 10.1136/bmj.324.7328.35

Berger, M., Yang, Q., & Maier, A. (2018). X-ray Imaging. In A. Maier, S. Steidl, V. Christlein, & J. Hornegger (Eds.), Medical Imaging Systems: An Introductory Guide. Springer. http://www.ncbi.nlm.nih.gov/books/NBK546155/

Bergmann, R. (2021, February 25). Getting Through an MRI When You Have Claustrophobia. Intermountain Medical Imaging. https://www.aboutimi.com/getting-through-an-mri-when-you-have-claustrophobia

Bianek-Bodzak, A., Szurowska, E., Sawicki, S., & Liro, M. (2013). The importance and perspective of magnetic resonance imaging in the evaluation of endometriosis. BioMed Research International, 2013. https://doi.org/10.1155/2013/436589

Biederer J. (2005). Magnetresonanztomographie-technische Grundlagen und aktuelle Entwicklungen [Magnetic resonance imaging: technical aspects and recent developments]. Medizinische Klinik (Munich, Germany : 1983), 100(1), 62–72. https://doi.org/10.1007/s00063-005-1124-z

Bloch, F. (1946). Nuclear Induction. Physical Review, 70(7-8), 460-474. https://doi.org/10.1103/PhysRev.70.460

Bonekamp, D., Jacobs, M. A., El-Khouli, R., Stoianovici, D., & Macura, K. J. (2011). Advancements in MR imaging of the prostate: from diagnosis to interventions. Radiographics, 31(3), 677–703. https://doi.org/10.1148/rg.313105139

Bourgioti, C., Preza, O., Panourgias, E., Chatoupis, K., Antoniou, A., Nikolaidou, M. E., & Moulopoulos, L. A. (2017). MR imaging of endometriosis: spectrum of disease. Diagnostic and Interventional Imaging, 98(11), 751-767. https://doi.org/10.1016/j.diii.2017.05.009

Breitzman, Anthony, "An Empirical Look at the Controversy Surrounding the Nobel Prize for Magnetic Resonance Imaging" (2017). Faculty Scholarship for the College of Science & Mathematics. 80. Retrieved from: https://rdw.rowan.edu/csm_facpub/80

Brice, "MRI Pioneers win Nobel Prize," Diagnostic Imaging Online, October 7, 2003.

Bringing More Value to Imaging Departments With MRI. (2019, June 10). Imaging Technology News. https://www.itnonline.com/article/bringing-more-value-imaging-departments-mri

Brown, M. A., & Semelka, R. C. (2003). MRI BASIC PRINCIPLES AND APPLICATIONS (third ed.). Wiley-Liss.

Buchanan, C. L., Morris, E. A., Dorn, P. L., Borgen, P. I., & Van Zee, K. J. (2005). Utility of Breast Magnetic Resonance Imaging in Patients With Occult Primary Breast Cancer. Annals of Surgical Oncology, 12(12), 1045–1053. https://doi.org/10.1245/aso.2005.03.520

Budjan, J., Schoenberg, S. O., Morelli, J. N., & Haneder, S. (2014). MR Contrast Agent Safety in the Age of Nephrogenic Systemic Fibrosis: Update 2014. Current Radiology Reports, 2(9). https://doi.org/10.1007/s40134-014-0064-x.

Bulas, D., & Egloff, A. (2013). Benefits and risks of MRI in pregnancy. Seminars in Perinatology, 37(5), 301–304. https://doi.org/10.1053/j.semperi.2013.06.005.

Bull, J. (1980). The History of Computed Tomography. In J.-M. Caillé & G. Salamon (Eds.), Computerized Tomography (pp. 3–6). Springer. https://doi.org/10.1007/978-3-642-67513-3_1

Bullmore, E. (2012). The future of functional MRI in clinical medicine. Neuroimage, 62(2), 1267–1271. https://doi.org/10.1016/j.neuroimage.2012.01.026

Burke, L. M. B., Ramalho, M., AlObaidy, M., Chang, E., Jay, M., & Semelka, R. C. (2016, May 19). Self-reported gadolinium toxicity: A survey of patients with chronic symptoms. Magnetic Resonance Imaging. https://www.sciencedirect.com/science/article/abs/pii/S0730725X16300510.

Callaghan, M. F., Negus, C., Leff, A. P., Creasey, M., Burns, S., Glensman, J., Bradbury, D., Williams, E., & Weiskopf. N. (2019). Safety of Tattoos in Persons Undergoing MRI. The New England Journal of Medicine. DOI: 10.1056/NEJMc1811197

Campbell, S. (2013). A Short History of Sonography in Obstetrics and Gynaecology. Facts, Views & Vision in ObGyn, 5(3), 213–229.

Canon Medical Systems. (2015, November 29). Pulmonary MR Imaging Is Now Possible with Toshiba's Ultrashort Echo Time Sequence. Press Releases. https://us.medical.canon/news/press-releases/2015/11/29/2263/

Caravan, P., Ellison, J. J., McMurry, T. J., & Lauffer, R. B. (1999, August 20). Gadolinium(III) Chelates as MRI Contrast Agents: Structure, Dynamics, and Applications. Chemical Reviews. https://pubs.acs.org/doi/10.1021/cr980440x.

Cardis, E., & Hatch, M. (2011, March 10). The Chernobyl Accident - An Epidemiological Perspective. Clinical Oncology. https://www.sciencedirect.com/science/article/abs/pii/S0936655511005425?casa_token=cq-gYZHuGOMAAAAA%3Ahcp06CTg8oJLlXu04V8I8Mp8y-GajUbyjotckf2ZWsmX-6551n7iv2MIOVHWb2dUFfQBEFteq2PU.

Carroll, P. R., Coakley, F. V., & Kurhanewicz, J. (2006). Magnetic resonance imaging and spectroscopy of prostate cancer. Reviews in Urology, 8(Suppl 1), S4–S10.

Castro, J. (2011, June 21). Brain Scans Predict Pop Hits. Scientific American. https://www.scientificamerican.com/article/brain-scans-predict-pop-hits/.

CBS. (2017, November 10). Chuck Norris and wife's lawsuit sparks debate over risks of MRI contrast agents. CBS News. https://www.cbsnews.com/news/chuck-norris-gena-norris-lawsuit-medical-companies-mri-contrast-agent/.

Chang, Y. C., Huang, K. M., Chen, J. H., & Su, C. T. (1999). Impact of magnetic resonance imaging on the advancement of medicine. Journal of the Formosan Medical Association = Taiwan Yi Zhi, 98(11), 740–748.

Chang, Y. C., Huang, K. M., Chen, J. H., & Su, C. T. (1999). Impact of magnetic resonance imaging on the advancement of medicine. Journal of the Formosan Medical Association, 98(11), 740–748.

Chen, H., Rogalski, M. M., & Anker, J. N. (2012). Advances in functional X-ray imaging techniques and contrast agents. Physical Chemistry Chemical Physics : PCCP, 14(39), 13469–13486. https://doi.org/10.1039/c2cp41858d

Chen, R., Jiao, Y., & Herskovits, E. H. (2011). Structural MRI in autism spectrum disorder. Pediatric Research, 69(8), 63–68. https://doi.org/10.1203/PDR.0b013e318212c2b3

Chhor, C. M., & Mercado, C. L. (2017). Abbreviated MRI protocols: wave of the future for breast cancer screening. American Journal of Roentgenology, 208(2), 284–289. https://doi.org/10.2214/AJR.16.17205

Chilla, G. S., Tan, C. H., Xu, C., & Poh, C. L. (2015). Diffusion weighted magnetic resonance imaging and its recent trend—a survey. Quantitative Imaging in Medicine and Surgery, 5(3), 407. https://doi.org/10.3978/j.issn.2223-4292.2015.03.01

COCIR. (n.d.). Medical Imaging. Retrieved May 08, 2021, from https://www.cocir.org/our-industry/medical-imaging.html

Constantinides, C. (2014). Fundamentals of magnetic resonance II: Imaging. Magnetic resonance imaging: The basics (pp. 39-51). CRC Press.

Corea, J. R., Flynn, A. M., Lechêne, B., Scott, G., Reed, G. D., Shin, P. J., Lustig, M., & Arias, A. C. (2016). Screen-printed flexible MRI receive coils. Nature Communications, 7(1). https://doi.org/10.1038/ncomms10839

Cosgrove, C. (2013, November 7). To relax during MRI, kids watch videos on Cinemavision goggles. MultiCare. https://www.multicare.org/news/to-relax-during-mri-kids-watch-videos-cinemavision-goggles/.

Das, B. K. (2015). Development of Positron Emission Tomography (PET): A Historical Perspective. In B. K. Das (Ed.), Positron Emission Tomography: A Guide for Clinicians (pp. 7–11). Springer India. https://doi.org/10.1007/978-81-322-2098-5_2

Davidovitch, M., Patterson, B., & Gartside, P. (1996). Head circumference measurements in children with autism. Journal of Child Neurology, 11(5), 389–393. https://doi.org/10.1177/088307389601100509

Dewey, M., Schink, T., & Dewey, C. F. (2007). Claustrophobia during magnetic resonance imaging: cohort study in over 55,000 patients. J Magn Reson Imaging, 26(5), 1322–1327. doi: 10.1002/jmri.21147. PMID: 17969166.

Dezube, R. (2019, July). Chest Imaging. Lung and Airway Disorders. https://www.merckmanuals.com/en-ca/home/lung-and-airway-disorders/diagnosis-of-lung-disorders/chest-imaging

Dichter, G. S. (2012). Functional magnetic resonance imaging of autism spectrum disorders. Dialogues in Clinical Neuroscience, 14(3), 319. https://doi.org/10.31887/DCNS.2012.14.3/gdichter

DMS Health. (2019, May 17). How Strong Does Your MRI Magnet Really Need to Be? https://www.dmshealth.com/how-strong-does-mri-magnet-need-to-be/.

Dournes, G., Walkup, L. L., Benlala, I., Willmering, M. M., Macey, J., Bui, S., Laurent, F., & Woods, J. C. (2020). The Clinical Use of Lung MRI in Cystic Fibrosis: What, Now, How? Chest, S0012-3692(20), 35453-2. Advance online publication. 10.1016/j.chest.2020.12.008

Durbridge, G. (2011). Magnetic Resonance Imaging: Fundamental Safety Issues. Journal of Orthopaedic & Sports Physical Therapy, 41(11), 820–828. https://doi.org/10.2519/jospt.2011.3906.

Durbridge, G. (2011). Magnetic Resonance Imaging: Fundamental Safety
     Issues. Journal of Orthopaedic & Sports Physical Therapy, 41(11):
     820–828.https://doi.org/10.2519/jospt.2011.3906

Edelman, R. R. (2014). The History of MR Imaging as Seen through the
     Pages of Radiology. Radiology, 273(2S), S181–S200. https://doi.
     org/10.1148/radiol.14140706

Edwards, A. D., & Arthurs, O. J.(2011).Paediatric MRI under sedation:is it nec-
     essary?What is the evidence for the alternatives?Pediatric Radiology,
     41(11),1353–1364. https://doi.org/10.1007/s00247-011-2147-7

Elster, A. D. (2021). Spin-warp imaging. Questions and Answers in MRI.
     http://mriquestions.com/spin-warp-imaging.html.

Emerson, R. W., Adams, C., Nishino, T., Hazlett, H. C., Wolff, J. J., Zwaigen-
     baum, L., & IBIS Network. (2017). Functional neuroimaging of
     high-risk 6-month-old infants predicts a diagnosis of autism at 24
     months of age. Science Translational Medicine, 9(393). https://doi.
     org/10.1126/scitranslmed.aag2882

Enders, J., Zimmermann, E., Rief, M., Martus, P., Klingebiel, R., Asbach, P.,
     Klessen, C., Diederichs, G., Bengner, T., Teichgräber, U., Hamm, B.,
     & Dewey, M. (2011). Reduction of claustrophobia during magnetic
     resonance imaging: methods and design of the "CLAUSTRO" ran-
     domized controlled trial. BMC medical imaging, 11, 4. https://doi.
     org/10.1186/1471-2342-11-4

FDA. (2017, December 9). Benefits and Risks. U.S. Food and Drug Adminis-
     tration. https://www.fda.gov/radiation-emitting-products/mri-mag-
     netic-resonance-imaging/benefits-and-risks#:~:text=MRI%20pro-
     vides%20better%20soft%20tissue,variety%20of%20diseases%20
     and%20conditions.

Felmlee J. P. (2005). The noise of MRI. Journal of the American College of Ra-
     diology : JACR, 2(6), 547. https://doi-org.myaccess.library.utoronto.
     ca/10.1016/j.jacr.2005.02.016.

Ferreira, A. M., Costa, F., Tralhão, A., Marques, H., Cardim, N., & Adragão, P.
     (2014). MRI-conditional pacemakers: current perspectives. Medi-
     cal devices (Auckland, N.Z.), 7, 115–124. https://doi.org/10.2147/
     MDER.S44063

Fields, E. C., & Weiss, E. (2016, February 2). A practical review of magnetic resonance imaging for the evaluation and management of cervical cancer. Radiation Oncology. https://ro-journal.biomedcentral.com/articles/10.1186/s13014-016-0591-0.

Filippi, M., Preziosa, P., & Rocca, M. A. (2018). MRI in multiple sclerosis: what is changing? Current Opinion in Neurology, 31(4), 386–395. https://doi.org/10.1097/wco.0000000000000572

Fishman, E. K., & Urban, B. A. (2001). Computed Tomography. In Cancer: Principles and Practice of Oncology (6th ed., Vol. 1). Lippincott Williams & Wilkins. https://oncouasd.files.wordpress.com/2014/09/cancer-principles-and-practice-of-oncology-6e.pdf

Földes, Z., Ala-Ruona, E., Burger, B., & Orsi, G. (2017). Anxiety reduction with music and tempo synchronization on magnetic resonance imaging patients. Psychomusicology: Music, Mind, and Brain, 27(4), 343–349. https://doi.org/10.1037/pmu0000199

Fonar Corporation. (2021a). FONAR's Innovation Timeline. Retrieved from Fonar Corporations: http://www.fonar.com/innovations-timeline.html

Fonar Corporation. (2021b). Our History . Retrieved from Fonar Corporation : http://www.fonar.com/history.html

Frank, A. (2016, September 22). The Freaky, Clanking, Buzzing, Whirring Glory Of MRI. NPR. https://www.npr.org/sections/13.7/2016/09/22/494988501/the-freaky-clanking-buzzing-whirring-glory-of-mri.

Fratini, M., Moraschi, M., Maraviglia, B., & Giove, F. (2014). On the impact of physiological noise in spinal cord functional MRI. Journal of magnetic resonance imaging : JMRI, 40(4), 770-777. National Center for Biotechnology Information. 10.1002/jmri.24467

Fraum, T. J., Ludwig, D. R., Bashir, M. R., & Fowler, K. J. (2017). Gadolinium-based contrast agents: A comprehensive risk assessment. Journal of Magnetic Resonance Imaging, 46(2), 338–353. https://doi.org/10.1002/jmri.25625.

Fryback, D. G., & Thornbury, J. R. (1991). The efficacy of diagnostic imaging. Medical Decision Making, 11(2), 88–94. https://doi.org/10.1177/0272989X9101100203

Fujimoto, J. G., Pitris, C., Boppart, S. A., & Brezinski, M. E. (2000). Optical Coherence Tomography: An Emerging Technology for Biomedical Imaging and Optical Biopsy. Neoplasia (New York, N.Y.), 2(1–2), 9–25.

Fung, B. (2012, April 26). 6 Cool Things People Have Done Inside MRI Scanners. The Atlantic. https://www.theatlantic.com/health/archive/2012/04/6-cool-things-people-have-done-inside-mri-scanners/256416/.

Gambhir S.S., & Yaghoubi S.S. (Eds). (2010). Chapter 7: Physics, Instrumentation, and Methods for Imaging Reporter Gene Expression in Living Subjects. Molecular imaging with reporter genes. Cambridge University Press.

Gass, A., Rocca, M. A., Agosta, F., Ciccarelli, O., Chard, D., Valsasina, P., & MAGNIMS Study Group. (2015). MRI monitoring of pathological changes in the spinal cord in patients with multiple sclerosis. The Lancet Neurology, 14(4), 443–454. https://doi.org/10.1016/S1474-4422(14)70294-7

Gay, C. (2018, December 18). MRI at the Movies: Venom. Shields MRI Blog. https://info.shields.com/mri-at-the-movies-venom.

GE Healthcare. (2017, June 6). Making MRI MAGiC by Cutting Scan Time for Patients and Docs. American Journal of Neuroradiology Publishes MAGiC Results. https://www.gehealthcare.com/article/making-mri-magic-by-cutting-scan-time-for-patients-and-docs

GE Healthcare. (2019). MRI History. Retrieved from GE Healthcare: https://www.gehealthcare.com/feature-article/when-and-why-was-mri-invented

GE Healthcare. (2019). The effects of arthroplasty on mri. Retrieved May 08, 2021, from https://www.gehealthcare.com/article/the-effects-of-arthroplasty-on-mri

GE Healthcare. (2019). This cardiac software originating from a Stanford basement is now one of the top of Artificial Intelligence solutions

available. Health Management, 19(2). https://healthmanagement.org/
uploads/article_attachment/hm-v19-i2-web-this-cardiac-software-orig-
inating-from-a-stanfort-basements.pdf

GE Healthcare. (2019a, May 13). MRI bore sizes and benefits. https://www.
gehealthcare.co.uk/article/mri-bore-sizes-and-benefits.

GE Healthcare. (2019b, March 13). The isocenter and the importance of magnet
homogeneity. https://www.gehealthcare.com/feature-article/the-iso-
center-and-the-importance-of-magnet-homogeneity.

GE Healthcare. (2019c, March 13). The isocenter and the importance of magnet
homogeneity. https://www.gehealthcare.com/feature-article/the-iso-
center-and-the-importance-of-magnet-homogeneity.

Gennisson J.-L., Deffieux T., Fink M., and Tanter M. (2013). Ultrasound elastography: Principles and techniques. Diagnostic and Interventional Imaging.
94(5), 487-495. doi:10.1016/j.dii.2013.01.022.

Gervais, H., Belin, P., Boddaert, N., Leboyer, M., Coez, A., Barthelemy, C., & Zilbo-
vicius, M. (2004). Abnormal voice processing in autism: a fMRI study.
Nature Neuroscience, (7), 801–802. https://doi.org/10.1038/nn1291

Gintoft, A. (2013, September 12). Shhhhhhhhhh GE Healthcare's Revolutionary Silent Scan Technology takes MRI Noise from a Rock Concert to a
Whisper201. GE Healthcare. https://www.ge.com/news/press-releas-
es/shhhhhhhhhh-ge-healthcares-revolutionary-silent-scan-technolo-
gy-takes-mri-noise-rock.

Glenn, O. A. (2010). MR imaging of the fetal brain. Pediatric Radiology, 40(1),
68–81. https://doi.org/10.1007/s00247-009-1459-3

Glenn, O. A., & Barkovich, A. J. (2006). Magnetic resonance imaging of the fetal
brain and spine: an increasingly important tool in prenatal diagnosis,
Part 1. American Journal of Neuroradiology, 27(8), 1604–1611.

Glover, G. H. (2011). Overview of functional magnetic resonance imaging.
Neurosurgery Clinics, 22(2), 133–139. https://doi.org/10.1016/j.
nec.2010.11.001

Grist, T. M. (2019, October 1). The Next Chapter in MRI: Back to the Future? Radiology, 293(2), 394-395. https://doi.org/10.1148/radiol.2019192011

Grist, T. M. (2019). The Next Chapter in MRI: Back to the Future? Radiology, 293(2), 394–395. https://doi.org/10.1148/radiol.2019192011

Grover, V. P. B., Tognarelli, J. M., Crossey, M. M. E., Cox, I. J., Taylor-Robinson, S. D., & McPhail, M. J. W. (2015). Magnetic Resonance Imaging: Principles and Techniques: Lessons for Clinicians. Journal of Clinical and Experimental Hepatology, 5(3), 246–255. https://doi.org/10.1016/j.jceh.2015.08.001

Gruber, B., Froeling, M., Leiner, T., & Klomp, D. W. J. (2018). RF coils: A practical guide for nonphysicists. Journal of Magnetic Resonance Imaging, 48(3), 590–604. https://doi.org/10.1002/jmri.26187

Gutierrez, C. (2011). Imaging Autism With MRI, Radiology Today. Retrieved May 4, 2020, from https://www.radiologytoday.net/archive/rt0211p6.shtml

Haase, S., & Maier, A. (2018). Endoscopy. In A. Maier, S. Steidl, V. Christlein, & J. Hornegger (Eds.), Medical Imaging Systems: An Introductory Guide. Springer. http://www.ncbi.nlm.nih.gov/books/NBK546146/

Hargreaves, B. A., Worters, P. W., Pauly, K. B., Pauly, K, M., & Gold, G. E. (2011). Metal-Induced Artifacts in MRI. AJR, 197, 547–555. DOI:10.2214/AJR.11.7364

Harisinghani, M. G., O'Shea, A., & Weissleder, R. (2019). Advances in clinical MRI technology. Science Translational Medicine, 11(523). https://doi.org/10.1126/scitranslmed.aba2591

Harisinghani, M. G., O'Shea, A., & Weissleder, R. (2019). Advances in clinical MRI technology. Science Translational Medicine, 11(523). https://doi.org/10.1126/scitranslmed.aba2591.

Harris, L. M., Cumming, S. R., & Menzies, R. G. (2004). Predicting anxiety in magnetic resonance imaging scans. International journal of behavioral medicine, 11(1), 1–7. https://doi.org/10.1207/s15327558ijbm1101_1

Hartman, A.-R., Daniel, B. L., Kurian, A. W., Mills, M. A., Nowels, K. W., Dirbas, F. M., ... Plevritis, S. K. (2004). Breast magnetic resonance image screening and ductal lavage in women at high genetic risk for breast carcinoma. Cancer, 100(3), 479–489. https://doi.org/10.1002/cncr.11926

Harvard Health Publishing. (2013, March). Do CT scans cause cancer? https://www.health.harvard.edu/staying-healthy/do-ct-scans-cause-cancer#:~:text=How%20much%20is%20too%20much,up%20to%2025%20chest%20CTs.

Harvard Medical School. (2010, March). By the way, doctor: Are mri contrast agents harmful? Retrieved May 08, 2021, from https://www.health.harvard.edu/newsletter_article/are-mri-contrast-agents-harmful

Harvard Women's Health Watch. (2020, January 29). Radiation risk from medical imaging. Retrieved May, from https://www.health.harvard.edu/cancer/radiation-risk-from-medical-imaging

Hatano, K., Sekiya, Y., Araki, H., Sakai, M., Togawa, T., Narita, Y., ... Ito, H. (1999). Evaluation of the therapeutic effect of radiotherapy on cervical cancer using magnetic resonance imaging. International Journal of Radiation Oncology*Biology*Physics, 45(3), 639–644. https://doi.org/10.1016/s0360-3016(99)00228-x

Havsteen, I., Ohlhues, A., Madsen, K. H., Nybing, J. D., Christensen, H., & Christensen, A. (2017). Are Movement Artifacts in Magnetic Resonance Imaging a Real Problem?—A Narrative Review. Frontiers in Neurology, 8, 232. National Center for Biotechnology Information. 10.3389/fneur.2017.00232

Hazlett, H. C., Gu, H., Munsell, B. C., Kim, S. H., Styner, M., Wolff, J. J., & Piven, J. (2017). Early brain development in infants at high risk for autism spectrum disorder. Nature, 542(7641), 348–351. https://doi.org/10.1038/nature21369

Holdsworth, S. J., Macpherson, S. J., Yeom, K. W., Wintermark, M., & Zaharchuk G. (2018). Clinical Evaluation of Silent T1-Weighted MRI and Silent MR Angiography of the Brain. American Journal of Roentgenology, 210(2), 404–411. https://doi.org/10.2214/ajr.17.18247.

Honal, M., Leupold, J., Huff, S., Baumann, T., & Ludwig, U. (2010, March). Compensation of breathing motion artifacts for MRI with continuous ly moving table. Magnetic Resonance in Medicine, 63(3), 701-712. National Center for Biotechnology Information. 10.1002/mrm.22162

Hornak, J.P. (2020). Image artifacts. The basics of MRI. Interactive Learning Software. https://www.cis.rit.edu/htbooks/mri/chap-11/chap11. htm#:~:text=An%20image%20artifact%20is%20any,properties%20 of%20the%20human%20body

Hoshi, Y., & Yamada, Y. (2016). Overview of diffuse optical tomography and its clinical applications. Journal of Biomedical Optics, 21(9), 091312. https://doi.org/10.1117/1.JBO.21.9.091312

Hricak, H., Gatsonis, C., Coakley, F. V., Snyder, B., Reinhold, C., Schwartz, L. H., ... Mitchell, D. G. (2007). Early Invasive Cervical Cancer: CT and MR Imaging in Preoperative Evaluation—ACRIN/GOG Comparative Study of Diagnostic Performance and Interobserver Variability. Radiology, 245(2), 491–498. https://doi.org/10.1148/radiol.2452061983

Hsu, A. L., Khachikyan, I., & Stratton, P. (2010). Invasive and non-invasive methods for the diagnosis of endometriosis. Clinical Obstetrics and Gynecology, 53(2), 413. https://doi.org/10.1097/GRF.0b013e3181d-b7ce8

Hudson, M., Seppälä, K., Putkinen, V., Sun, L., Glerean, E., Karjalainen, T., & Nummenmaa, L. (2020, August 1). Dissociable neural systems for unconditioned acute and sustained fear. NeuroImage. https://www. sciencedirect.com/science/article/pii/S1053811920300094?via%-3Dihub.

Huisman, T. A. G. M. (2010). Diffusion-weighted and diffusion tensor imaging of the brain, made easy. Cancer Imaging, 10(1A), 163-171. https:// doi.org/10.1102/1470-7330.2010.9023

Ibrahim, M. A., Harzhirkarzar, B., & Dublin, A. B. (2020). Gadolinium Magnetic Resonance Imaging. StatPearls.

Ibrahim, N. M. A., & Elsaeed, H. H. (2012). The role of MRI in the diagnosis of endometriosis. The Egyptian Journal of Radiology and Nuclear Medicine, 43(4), 631–636. https://doi.org/10.1016/j.ejrnm.2012.09.005

Imaging Technology News. (2016, September 6). GE Gains FDA Clearance for Multi-Contrast MRI in a Single Acquisition. Magnetic Resonance

Imaging (MRI). https://www.itnonline.com/content/ge-gains-fda-clearance-multi-contrast-mri-single-acquisition

Imaging Technology News. (2019, August 27). Benefits of Wide Bore MRI. https://www.itnonline.com/article/benefits-wide-bore-mri#:~:text=Wide%20Bore%20Versus%20Conventional%20Open%20Bore&text=The%20introduction%20of%20Siemens'%20Espree,a%20brand%20new%20MRI%20technology.

Imaging Technology News. (2020, May 27). United Imaging Debuts Ultra-wide 75 cm Bore 3T MRI. https://www.itnonline.com/content/united-imaging-debuts-ultra-wide-75-cm-bore-3t-mri.

InsideRadiology. (2018, October 31). Magnetic Resonance Imaging (MRI). https://www.insideradiology.com.au/mri/.

Iwan, E., Yang, J., Enders, J., Napp, A. E., Rief, M., & Dewey, M. (2020). Patient preferences for development in MRI scanner design: a survey of claustrophobic patients in a randomized study. European Radiology, 31, 1325-1335.

J. (2020, June 23). Magnetic Resonance Imaging (MRI). Familydoctor.Org. https://familydoctor.org/magnetic-resonance-imaging-mri/

Jha, P., Sakala, M., Chamie, L. P., Feldman, M., Hindman, N., Huang, C., & Taffel, M. T. (2019). Endometriosis MRI lexicon: consensus statement from the society of abdominal radiology endometriosis disease-focused panel. Abdominal Radiology, 1–17. https://doi.org/10.1007/s00261-019-02291-x

Jim Stallard. (2019, May 10). CT vs MRI: What's the Difference? And How Do Doctors Choose Which Imaging Method to Use? Memorial Sloan Kettering Cancer Center. https://www.mskcc.org/news/ct-vs-mri-what-s-difference-and-how-do-doctors-choose-which-imaging-method-use.

Jin, J. (1999). Electromagnetic analysis and design in magnetic resonance imaging. Taylor & Francis Group.

Jin, J. (1999). Introduction to magnetic resonance imaging. Electromagnetic analysis and design in magnetic resonance imaging (pp. 1-37). CRC Press.

Jones, J. (2020). Larmor frequency: Radiology reference article. Radiopaedia Blog RSS. https://radiopaedia.org/articles/larmor-frequency?lang=us

Jones, T., & Townsend, D. (2017). History and future technical innovation in positron emission tomography. Journal of Medical Imaging, 4(1). https://doi.org/10.1117/1.JMI.4.1.011013

Jouanneau, E., Messerer, M., & Berhouma, M. (2011). Endoscopic Endonasal Skull Base Surgery: Current State of the Art and Future Trends. https://doi.org/10.5772/24768

Kadom, N., & Trofimova, A. (2019). GOBrain 5-Minute MRI in Children: Shown to Reduce the Need for Sedation. MAGNETOM Flash, (68). https://cdn0.scrvt.com/39b415fb07de4d9656c7b516d-8e2d907/1800000004387126/48b07be6d54e/magnetom-flash-69_gobrain-5-minute-mri-in-children_kadom_1800000004387126.pdf

Kasban, H., El-bendary, M., & Salama, D. (2015). A Comparative Study of Medical Imaging Techniques. International Journal of Information Science and Intelligent System, 4, 37–58.

Keilman, C., & Shanks, A. L. (2020). Oligohydramnios. StatPearls. Retrieved May 4, 2021, from https://www.ncbi.nlm.nih.gov/books/NBK562326/

Khalil M.M., Tremoleda J.L., Bayomy T.B., & Gsell W. (2011). Molecular SPECT imaging: an overview. International Journal of Molecular Imaging. 2011(796025): 1-15.

Kherlopian A.R., Song T., Duan Q., Neimark M.A., Po M.J., Gohagan J.K., & Laine A.F. (2008). A review of imaging techniques for systems biology. BMC Systems Biology. 2(74). Doi: 10.1186/1752-0509-2-74.

Kissane, J., Neutze, J. A., & Singh, H. (Eds.). (2020). Radiology Fundamentals-Introduction to Imaging and Technology (Sixth ed.). Springer Nature https://doi.org/10.1007/978-3-030-22173-7_7

Kleinhans, N. M., Müller, R. A., Cohen, D. N., & Courchesne, E. (2008). Atypical functional lateralization of language in autism spectrum disorders. Brain Research, 1221, 115–125. http://dx.doi.org/10.1016/j.brainres.2008.04.080

Kose, K. (2021). Physical and technical aspects of human magnetic resonance imaging: Present status and 50 years historical review. Advances in Physics: X, 6(1), 1885310. https://doi.org/10.1080/23746149.2021.1885310

Krans, B. (2018, September 29). Heart MRI: Purpose, Procedure, and Risks. Healthline. https://www.healthline.com/health/heart-mri.

Kriege, M., Brekelmans, C. T. M., Boetes, C., Besnard, P. E., Zonderland, H. M., Obdeijn, I. M., ... Klijn, J. G. M. (2004). Efficacy of MRI and Mammography for Breast-Cancer Screening in Women with a Familial or Genetic Predisposition. New England Journal of Medicine, 351(5), 427–437. https://doi.org/10.1056/nejmoa031759

Kriege, M., Brekelmans, C. T., Boetes, C., Besnard, P. E., Zonderland, H. M., Obdeijn, I. M., Manoliu, R. A., Kok, T., Peterse, H., Tilanus-Linthorst, M. M., Muller, S. H., Meijer, S., Oosterwijk, J. C., Beex, L. V., Tollenaar, R. A., de Koning, H. J., Rutgers, E. J., Klijn, J. G., & Magnetic Resonance Imaging Screening Study Group (2004). Efficacy of MRI and mammography for breast-cancer screening in women with a familial or genetic predisposition. The New England journal of medicine, 351(5), 427–437. https://doi.org/10.1056/NEJMoa031759.

Kuhl, C. K., Schrading, S., Strobel, K., Schild, H. H., Hilgers, R. D., & Bieling, H. B. (2014). Abbreviated breast magnetic resonance imaging (MRI): first postcontrast subtracted images and maximum-intensity projection—a novel approach to breast cancer screening with MRI. Journal of Clinical Oncology, 32(22), 2304–2310. https://doi.org/10.1200/JCO.2013.52.5386

Kumar, S., Liney, G., Rai, R., Holloway, L., Moses, D., & Vinod, S. K. (2016, April). Magnetic resonance imaging in lung: a review of its potential for radiotherapy. The British Journal of Radiology, 89(1060). National Center for Biotechnology Information. 10.1259/bjr.20150431

Kuo, P. H. (2008). NSF-Active and NSF-Inert Species of Gadolinium: Mechanistic and Clinical Implications. American Journal of Roentgenology, 191(6), 1861–1863. https://doi.org/10.2214/ajr.08.1179.

Lalonde, L., David, J., & Trop, I. (2005). Magnetic resonance imaging of the breast: current indications. Canadian Association of Radiologists, 56(5), 301–308.

Lauterbur, P. C. (2003). Paul C. Lauterbur - Biographical. Retrieved from The Nobel Prize : https://www.nobelprize.org/prizes/medicine/2003/lauterbur/biographical/

Lehman, C. D., Gatsonis, C., Kuhl, C. K., Hendrick, R. E., Pisano, E. D., Hanna, L., & Schnall, M. D. (2007). MRI evaluation of the contralateral breast in women with recently diagnosed breast cancer. New England Journal of Medicine, 356(13), 1295–1303. https://doi.org/10.1056/NEJ-Moa065447

Levine, D., Barnes, P. D., Madsen, J. R., Abbott, J., Mehta, T., & Edelman, R. R. (1999). Central nervous system abnormalities assessed with prenatal magnetic resonance imaging. Obstetrics & Gynecology, 94(6), 1011–1019. https://doi.org/10.1016/S0029-7844(99)00455-X

Lukins, R., Davan, I. G., & Drummond, P. D. (1997). A cognitive behavioural approach to preventing anxiety during magnetic resonance imaging. Journal of behavior therapy and experimental psychiatry, 28(2), 97–104. https://doi.org/10.1016/s0005-7916(97)00006-2

Manatt, S. (2013, June 3). Illuminating Facts About MRI. Chemical & Engineering News Magazine, 91(22). Retrieved from: https://cen.acs.org/articles/91/i22/Illuminating-Facts-MRI.html

Mann, R. M., Cho, N., & Moy, L. (2019). Breast MRI: state of the art. Radiology, 292(3), 520– 536. https://doi.org/10.1148/radiol.2019182947

Mansfield, S. P. (2003). Sir Peter Mansfield - Biographical . Retrieved from The Nobel Prize : https://www.nobelprize.org/prizes/medicine/2003/mansfield/biographical/

Marckmann, P., Skov, L., Rossen, K., Dupont, A., Damholt, M. B., Heaf, J. G., & Thomsen, H. S. (2006). Nephrogenic Systemic Fibrosis: Suspected Causative Role of Gadodiamide Used for Contrast-Enhanced Magnetic Resonance Imaging. Journal of the American Society of Nephrology, 17(9), 2359–2362. https://doi.org/10.1681/asn.2006060601.

Marcu, C. B., Beek, A. M., & van Rossum, A. C. (2006). Clinical applications of cardiovascular magnetic resonance imaging. Canadian Medical Association Journal, 175(8), 911–917. https://doi.org/10.1503/cmaj.060566

Mariappan, Y. K., Glaser, K. J., & Ehman, R. L. (2010). MAGNETIC RESONANCE ELASTOGRAPHY: A REVIEW. Clinical Anatomy (New York, N.Y.), 23(5), 497–511. https://doi.org/10.1002/ca.21006

Matsuo-Hagiyama, C., Watanabe, Y., Tanaka, H., Takahashi, H., Arisawa, A., Yoshioka, E., & Tomiyama, N. (2017). Comparison of Silent and Conventional MR Imaging for the Evaluation of Myelination in Children. Magnetic Resonance in Medical Sciences, 16(3), 209–216. https://doi.org/10.2463/mrms.mp.2016-0045.

Mayo Clinic. (2019, June 08). Nephrogenic systemic Fibrosis. Retrieved May 08, 2021, from https://www.mayoclinic.org/diseases-conditions/nephrogenic-systemic-fibrosis/symptoms-causes/syc-20352299

McAlonan, G. M., Cheung, V., Cheung, C., Suckling, J., Lam, G. Y., Tai, K. S., & Chua, S. E.(2005). Mapping the brain in autism. A voxel-based MRI study of volumetric differences and intercorrelations in autism. Brain, 128(2), 268–276. https://doi.org/10.1093/brain/awh332

McGregor, J., & Thompson, G. (2018). Role of MRI in uterine didelphys with co-existing endometrial carcinosarcoma. British Journal of Radiology | Case Reports, 4(4), 20180010. https://doi.org/10.1259/bjr-cr.20180010

McNamara, L. (2015, October 5). What is Multiple Sclerosis (MS)?: The Johns Hopkins Multiple Sclerosis Center. What is Multiple Sclerosis (MS)? | The Johns Hopkins Multiple Sclerosis Center. https://www.hopkinsmedicine.org/neurology_neurosurgery/centers_clinics/multiple_sclerosis/conditions/.

McNulty, J., & McNulty, S. (2009). Acoustic noise in magnetic resonance imaging: An ongoing issue. Radiography, 15(4), 320–326. https://doi.org/10.1016/j.radi.2009.01.001

Meier, D. S., Weiner, H. L., & Guttmann, C. R. (2007). MR imaging intensity modeling of damage and repair in multiple sclerosis: relationship of short-term lesion recovery to progression and disability. American Journal of Neuroradiology, 28(10), 1956–1963. https://doi.org/10.3174/ajnr.A0701

Merriam-Webster. (n.d.). Gyromagnetic ratio. In Merriam-Webster.com dictionary. Retrieved May 6, 2020, from https://www.merriam-webster.

com/dictionary/gyromagnetic%20ratio

Miller, E., & Smith, B. (2017). Case Series: Pediatric GOBrain-5-Minute Protocol MR Imaging at 3 Tesla. MAGNETOM Flash, (68). https://cdn0.scrvt.com/39b415fb07de4d9656c7b516d-8e2d907/1800000006165628/817a84f81af8/Case_Series_Pediatric_GOBrain-5-Minute_Protocol_MR_Imaging_at_3_Tesla_1800000006165628.pdf

Miller, G. W., Mugler, J. P., Sá, R. C., Altes, T. A., Prisk, G. K., & Hopkins, S. R. (2014). Advances in Functional and Structural Imaging of the Human Lung Using Proton MRI. NMR in Biomedicine, 27(12), 1542-1556. National Center for Biotechnology Information. 10.1002/nbm.3156

Millischer, A. E., Salomon, L. J., Santulli, P., Borghese, B., Dousset, B., & Chapron, C. (2015). Fusion imaging for evaluation of deep infiltrating endometriosis: feasibility and preliminary results. Ultrasound in Obstetrics & Gynecology, 46(1), 109–117. https://doi.org/10.1002/uog.14712

Moratal, D., Valles-Luch, A., Marti-Bonmati, L., & Brummer, M. E. (2008). k-Space tutorial: an MRI educational tool for a better understanding of k-space. Biomedical Imaging and Intervention Journal, 4(1). https://doi.org/10.2349/biij.4.1.e15

Morris, E. A., & Liberman, L. (2005). Breast MRI: diagnosis and intervention. Springer Nature. https://books.google.com/books?hl=en&lr=&id=5hpBzXTFpB8C&oi=fnd&pg=PR7&dq=breast+mri&ots=L-uhtl3zTk&sig=GslRmXisWBVRDa_tH8yLpsow_pg

MRI costs. (2018, October 30). Imaging Technology News. https://www.itnonline.com/content/mri-costs

MRI Frequently Asked Questions. (2021). MRI Frequently Asked Question. https://www.rmh.org/programs-and-services/mri-frequently-asked-questions

MRI Gadolinium Toxicity. (2012). MRI Gadolinium Toxicity Support Group. RSS. https://mri-gadolinium-toxicity.groups.io/g/Support.

MRI Safety During Pregnancy. (2019). Patient Safety - MRI During Pregnancy - RadiologyInfo.Org. https://www.radiologyinfo.org/en/info/

safety-mri-pregnancy#:~:text=There%20are%20no%20proven%20
risks,the%20baby%20have%20been%20found.

Müller, R. A., Behen, M. E., Rothermel, R. D., Chugani, D. C., Muzik, O., Mangner, T. J., & Chugani, H. T. (1999). Brain mapping of language and auditory perception in high-functioning autistic adults: a PET study. Journal of Autism and Developmental Disorders, 29(1), 19–31. https://doi. org/10.1023/A:1025914515203

Munn, Z., & Jordan, Z. (2013). Interventions to reduce anxiety, distress and the need for sedation in adult patients undergoing magnetic resonance imaging: a systematic review. Int J Evid Based Healthc, 11(4), 265–274.

Munn, Z., Moola, S., Lisy, K., Riitano, D., & Murphy, F. (2015). Claustrophobia in magnetic resonance imaging: A systematic review and meta-analysis. Radiography, 21(2). DOI:https://doi.org/10.1016/j.radi.2014.12.004

Munn, Z., Pearson, A., Jordan, Z., Murphy, F., Pilkington, D., & Anderson, A. (2015). Patient Anxiety and Satisfaction in a Magnetic Resonance Imaging Department: Initial Results from an Action Research Study. Journal of medical imaging and radiation sciences, 46(1), 23–29. https://doi.org/10.1016/j.jmir.2014.07.006

Murphy, A. (2020). Motion artifact. Radiology Reference Article. https://ra-diopaedia.org/articles/motion-artifact-2

Murphy, A., & Ballinger, J. R. (n.d.). Magnets (types) | Radiology Reference Article | Radiopaedia.org. Radiopaedia. https://radiopaedia.org/arti-cles/magnets-types

Murphy, G., Haider, M., Ghai, S., & Sreeharsha, B. (2013). The expanding role of MRI in prostate cancer. American Journal of Roentgenology, 201(6), 1229–1238. http://dx.doi.org/10.2214/AJR.12.10178

Nandakumar, G., & Fleshman, J. W. (2010). Laparoscopy for Colon and Rectal Cancer. Clinics in Colon and Rectal Surgery, 23(1), 51–58. https://doi. org/10.1055/s-0030-1247856

Natarajan, S., Marks, L. S., Margolis, D. J., Huang, J., Macairan, M. L., Lieu, P., & Fenster, A. (2011). Clinical application of a 3D ultrasound-guided prostate biopsy system. Urologic Oncology: Seminars and Original

Investigations, 29(3), 334–342. https://doi.org/10.1016/j.uro-lonc.2011.02.014

National Institute of Biomedical Imaging and Bioengineering. (2016, July). Ultrasound. https://www.nibib.nih.gov/science-education/science-topics/ultrasound

National Institute of Biomedical Imaging and Bioengineering. (n.d.). Computed Tomography (CT). https://www.nibib.nih.gov/science-education/science-topics/computed-tomography-ct

National Institute of Biomedical Imaging and Bioengineering. (n.d.). Magnetic resonance imaging (MRI). https://www.nibib.nih.gov/science-education/science-topics/magnetic-resonance-imaging-mri

National Institute of Biomedical Imaging and Bioengineering. (n.d.). Magnetic Resonance Imaging (MRI). https://www.nibib.nih.gov/science-education/science-topics/magnetic-resonance-imaging-mri.

National Institute of Neurological Disorders and Stroke. (2020, April). Hydrocephalus Fact Sheet. https://www.ninds.nih.gov/Disorders/Patient-Caregiver-Education/Fact-Sheets/Hydrocephalus-Fact-Sheet

National Maximum Wait time Access Targets for Medical Imaging. (2013). MRI & CT - Canadian Association of Radiologists. https://car.ca/wp-content/uploads/car-national-maximum-waittime-targets-mri-and-ct.pdf

Nave, C.R. (2017a). Magnetic dipole moment. HyperPhysics. http://hyperphysics.phy-astr.gsu.edu/hbase/magnetic/magmom.html

Nave, C.R. (2017b). Nuclear Spin. HyperPhysics. http://hyperphysics.phy-astr.gsu.edu/hbase/Nuclear/nspin.html#c2

NIBIB. (2018, April 3). Magnetic Resonance Imaging (MRI). National Institute of Biomedical Imaging and Bioengineering. https://www.nibib.nih.gov/science-education/science-topics/magnetic-resonance-imaging-mri.

Nicholson, J (2018). How Does a Solenoid Work? Sciencing. https://sciencing.com/a-solenoid-work-4567178.html

NINDS. (2020, August 1). Brain and Spinal Cord Tumors: Hope Through Research. National Institute of Neurological Disorders and Stroke. https://www.ninds.nih.gov/Disorders/Patient-Caregiver-Education/ Hope-Through-Research/Brain-and-Spinal-Tumors-Hope-Through.

Niwa, H. (2008). The History of Digestive Endoscopy. In H. Niwa, H. Tajiri, M. Nakajima, & K. Yasuda (Eds.), New Challenges in Gastrointestinal Endoscopy (pp. 3–28). Springer Japan. https://doi.org/10.1007/978-4-431-78889-8_1

O'Connor, M. (2020, January 28). What MRI and 100 years of horror movies can teach us about our brains. Health Imaging. https://www.health-imaging.com/topics/advanced-visualization/mri-horror-movies-tell-us-about-brains-pet.

Ohlmann-Knafo, S., Morlo, M., Tarnoki, D. L., Tarnoki, A. D., Grabowski, B., Kaspar, M., & Pickuth, D. (2016). Comparison of image quality characteristics on Silent MR versus conventional MR imaging of brain lesions at 3 Tesla. The British Journal of Radiology, 89(1067), 20150801. https://doi.org/10.1259/bjr.20150801.

Okinawa Institute of Science and Technology. (2020, April). Scientists unveil how general anesthesia works. News Center. https://www.oist.jp/ news-center/news/2020/4/24/scientists-unveil-how-general-anesthesia-works

Optical Imaging. (n.d.). Retrieved May 7, 2021, from https://www.nibib.nih. gov/science-education/science-topics/optical-imaging

Pagani, E., Bizzi, A., Di Salle, F., De Stefano, N., & Filippi, M. (2008). Basic concepts of advanced MRI techniques. Neurological Sciences, 29(S3), 290–295. https://doi.org/10.1007/s10072-008-1001-7

Palmer, W. (2020, May 26). First Ultra-Wide Bore MRI Receives FDA Clearance. Diagnostic Imaging. https://www.diagnosticimaging.com/ view/first-ultra-wide-bore-mri-receives-fda-clearance.

Park, B. K., Park, J. W., Park, S. Y., Kim, C. K., Lee, H. M., Jeon, S. S., & Choi, H. Y. (2011). Prospective evaluation of 3-T MRI performed before initial transrectal ultrasound–guided prostate biopsy in patients with high prostate-specific antigen and no previous biopsy. American Journal of Roentgenology, 197(5), W876–W881. http://dx.doi.org/ 10.2214/ AJR.11.6829

Partain, L. (2004). The 2003 Nobel prize for MRI: Significance and impact. Journal of Magnetic Resonance Imaging, 19(5), 515-526. doi:https:// doi.org/10.1002/jmri.20035

Patterson, K. (2017, November 2). Chuck Norris Claims MRI Chemical Poisoned His Wife. Popculture. https://popculture.com/celebrity/news/ chuck-norris-wife-mri-gadolinium-poisoning/.

Paulson, E. S., Erickson, B., Schultz, C., & Allen Li, X. (2014). Comprehensive MRI simulation methodology using a dedicated MRI scanner in radiation oncology for external beam radiation treatment planning. Medical Physics, 42(1), 28–39. https://doi.org/10.1118/1.4896096

Pinto, P. A., Chung, P. H., Rastinehad, A. R., Baccala, A. A., Kruecker, J., Benjamin, C. J., & Wood, B. J. (2011). Magnetic resonance imaging/ultrasound fusion guided prostate biopsy improves cancer detection following transrectal ultrasound biopsy and correlates with multiparametric magnetic resonance imaging. The Journal of Urology, 186(4), 1281–1285. https://doi.org/10.1016/j.juro.2011.05.078

Piven, J., Bailey, J., Ranson, B. J., & Arndt, S. (1997). An MRI study of the corpus callosum in autism. American Journal of Psychiatry, 154(8), 1051–1056. https://doi.org/10.1176/ajp.154.8.1051

POM. (2016, August 15). Are MRIs the Best Tool for Early Cancer Detection? POM Blog. https://www.pommri.com/blog/are-mris-the-best-tool-for-early-cancer-detection/.

Porcaro, A. B., Borsato, A., Romano, M., Sava, T., Ghimenton, C., Migliorini, F., & Montemezzi, S. (2013). Accuracy of preoperative endo-rectal coil magnetic resonance imaging in detecting clinical under-staging of localized prostate cancer. World Journal of Urology, 31(5), 1245–1251. https://doi.org/10.1007/s00345-012-0900-7

Portnow, L. H., Vaillancourt, D. E., & Okun, M. S. (2013). The history of cerebral PET scanning. Neurology, 80(10), 952–956. https://doi. org/10.1212/WNL.0b013e318285c135

Pötter, R., Georg, P., Dimopoulos, J. C. A., Grimm, M., Berger, D., Nesvacil, N., ...

Kirisits, C. (2011). Clinical outcome of protocol based image (MRI) guided adaptive brachytherapy combined with 3D conformal radiotherapy with or without chemotherapy in patients with locally advanced cervical cancer. Radiotherapy and Oncology, 100(1), 116–123. https://doi.org/10.1016/j.radonc.2011.07.012

Preston, D.C. (2016). Basics of MRI. Neuroimaging in neurology: An interactive approach. https://case.edu/med/neurology/NR/MRI%20Basics.htm

Purves D., Augustine G.J., Fitzpatrick D., Hall W.C., LaManita AS., Mooney R.D., Platt M.L., & White L.E. (Eds.). (2018). Chapter 1: Studying the Nervous System. Neuroscience (6th ed.). Oxford University Press.

Quirk, M. E., Letendre, A. J., Ciottone, R. A., & Lingley, J. F. (1989). Anxiety in patients undergoing MR imaging. Radiology, 170(2), 463–466. https://doi.org/10.1148/radiology.170.2.2911670

Quirk, M. E., Letendre, A. J., Ciottone, R. A., & Lingley, J. F. (1989). Evaluation of three psychologic interventions to reduce anxiety during MR imaging. Radiology, 173(3), 759–762. https://doi.org/10.1148/radiology.173.3.2682775

RAI. (2019, February 26). Is it safe to have an Mri scan with Tattoos? Here is what you need to know. Retrieved May 08, 2021, from https://4rai.com/blog/is-it-safe-to-have-an-mri-scan-with-tattoos-here-is-what-you-need-to-know

Raikhlin, A., Curpen, B., Warner, E., Betel, C., Wright, B., & Jong, R. (2015). Breast MRI as an adjunct to mammography for breast cancer screening in high-risk patients: retrospective review. American Journal of Roentgenology, 204(4), 889–897. https://www.ajronline.org/doi/10.2214/AJR.13.12264

Raplino, O., & Heberlein, K. (2016). New Strategies for Protocol Optimization for Clinical MRI: Rapid Examinations and Improved Patient Care. MAGNETOM Flash, (65). https://cdn0.scrvt.com/39b415fb07d-e4d9656c7b516d8e2d907/1800000003051764/63161f2dfba5/New-Strategies-for-Protocol-Optimization-for-Clinical-MR-Rapid-Examinations-and-Improved-Patient-Care_1800000003051764.pdf

Ray, C. C. (2012, April 16). The Sound and the Fury. The New York Times. https://www.nytimes.com/2012/04/17/science/why-mri-machines-make-that-loud-noise.html.

Ray, J. G., Vermeulen, M. J., Bharatha, A., Montanera, W. J., & Park, A. L. (2016). Association Between MRI Exposure During Pregnancy and Fetal and Childhood Outcomes. JAMA, 316(9), 952. https://doi.org/10.1001/jama.2016.12126.

Redcay, E., & Courchesne, E. (2008). Deviant functional magnetic resonance imaging patterns of brain activity to speech in 2–3-year-old children with autism spectrum disorder. Biological Psychiatry, 64(7), 589–598. https://doi.org/10.1016/j.biopsych.2008.05.020

Reed, E. (2019, May 29). How Much Does an MRI Cost? TheStreet. https://www.thestreet.com/lifestyle/health/how-much-does-an-mri-cost-14972340.

Rey, J. W., Kiesslich, R., & Hoffman, A. (2014). New aspects of modern endoscopy. World Journal of Gastrointestinal Endoscopy, 6(8), 334–344. https://doi.org/10.4253/wjge.v6.i8.334

Rich, D. A. (1997). A Brief History of Positron Emission Tomography. Journal of Nuclear Medicine Technology, 25(1), 4–11.

Righart, R., Biberacher, V., Jonkman, L. E., Klaver, R., Schmidt, P., Buck, D., & Mühlau, M. (2017). Cortical pathology in multiple sclerosis detected by the T1/T2-weighted ratio from routine magnetic resonance imaging. Annals of Neurology, 82(4), 519–529. https://doi.org/10.1002/ana.25020

Risk Factors. (2021). Stanford Health Care. https://stanfordhealthcare.org/medical-tests/m/mri/risk-factors.html

Roach, D. J., Crémillieux, Y., Fleck, R. J., Brody, A. S., Serai, S. D., Szczesniak, R. D., Kerlakian, S., Clancy, J. P., & Woods, J. C. (2016, November). Ultrashort Echo-Time Magnetic Resonance Imaging Is a Sensitive Method for the Evaluation of Early Cystic Fibrosis Lung Disease. Annals of the American Thoracic Society, 13(11), 1923-1931. National Center for Biotechnology Information. 10.1513/AnnalsATS.201603-203OC

RSNA. (2018, August 10). Magnetic Resonance Imaging (MRI) - Cardiac (Heart). Radiologyinfo.org. https://www.radiologyinfo.org/en/info/cardiacmr.

RSNA. (2018, June 22). Cervical Cancer. Radiologyinfo.org. https://www.radiologyinfo.org/en/info/cervicalcancer.

Saleem, S. N. (2014). Fetal MRI: An approach to practice: A review. Journal of Advanced Research, 5(5), 507–523. https://doi.org/10.1016/j.jare.2013.06.001

Sammet, S. (2016). Magnetic Resonance Safety. Abdominal Radiology (New York), 41(3), 444–451. https://doi.org/10.1007/s00261-016-0680-4

Schieda, N., Blaichman, J. I., Costa, A. F., Glikstein, R., Hurrell, C., James, M., Jabehdar Maralani, P., Shabana, W., Tang, A., Tsampalieros, A., van der Pol, C. B., & Hiremath, S. (2018). Gadolinium-Based Contrast Agents in Kidney Disease: A Comprehensive Review and Clinical Practice Guideline Issued by the Canadian Association of Radiologists. Canadian journal of kidney health and disease, 5, 2054358118778573. https://doi.org/10.1177/2054358118778573

Semelka, R. C., Ramalho, J., Vakharia, A., AlObaidy, M., Burke, L. M., Jay, M., & Ramalho, M. (2016, August 13). Gadolinium deposition disease: Initial description of a disease that has been around for a while. Magnetic Resonance Imaging. https://www.sciencedirect.com/science/article/abs/pii/S0730725X16301035?casa_token=6YYPD-kut3H8AAAAA%3ArspJ7FSRNIu_Cdh8jaCwVmt5NaR1Rc2YUi6pvHn-nIQTuROU3ArYb7hcT1Ngi8llbrKZjvkAcVmE.

Sepanlou, S. G., Safiri, S., Bisignano, C., Ikuta, K. S., Merat, S., Saberifiroozi, M., … Malekzadeh, R. (2020). The global, regional, and national burden of cirrhosis by cause in 195 countries and territories, 1990–2017: a systematic analysis for the Global Burden of Disease Study 2017. The Lancet Gastroenterology & Hepatology, 5(3), 245–266. https://doi.org/10.1016/s2468-1253(19)30349-8

Shampo, M. A., Kyle, R. A., & Steensma, D. P. (2012, February). Isidor Rabi—1944 Nobel Laureate in Physics. In Mayo Clinic Proceedings (Vol. 87, No. 2, p. e11). Elsevier.

Shellock, F. G., Woods, T. O., & Crues, J. V. (2009). MR Labeling Information for

Implants and Devices: Explanation of Terminology. Radiology, 253(1), 26–30. doi: 10.1148/radiol.2531091030

Siegelman, E. S., & Oliver, E. R. (2012). MR imaging of endometriosis: ten imaging pearls. Radiographics, 32(6), 1675–1691. https://doi.org/10.1148/rg.326125518

Sigrist, R. M. S., Liau, J., Kaffas, A. E., Chammas, M. C., & Willmann, J. K. (2017). Ultrasound Elastography: Review of Techniques and Clinical Applications. Theranostics, 7(5), 1303–1329. https://doi.org/10.7150/thno.18650

Singh, S. P. (2014). Magnetoencephalography: Basic principles. Annals of Indian Academy of Neurology, 17(Suppl 1), S107–S112. https://doi.org/10.4103/0972-2327.128676

Sivasubramanian, M., Hsia, Y., & Lo, L.-W. (2014). Nanoparticle-facilitated Functional and Molecular Imaging for the Early Detection of Cancer. Frontiers in Molecular Biosciences, 1. https://doi.org/10.3389/fmolb.2014.00015

Sobol, W. T. (2021, August 11). Recent advances in MRI technology: Implications for image quality and patient safety. Saudi Journal of Ophthalmology, 26(4), 393-399. National Center for Biotechnology Information. 10.1016/j.sjopt.2012.07.005

Sofka, C. M., Potter, H. G., Figgie, M., & Laskin, R. (2003). Magnetic Resonance Imaging of Total Knee Arthroplasty. Clinical Orthopaedics and Related Research, 406(1), 129–135.

Spaner, S. J., & Warnock, G. L. (1997). A brief history of endoscopy, laparoscopy, and laparoscopic surgery. Journal of Laparoendoscopic & Advanced Surgical Techniques. Part A, 7(6), 369–373. https://doi.org/10.1089/lap.1997.7.369

Sperling, J. W., Potter, H. G., Craig, E. V., Flatow, E., & Warren, R. F. (2002). MRI of the painful shoulder arthroplasty. J Shoulder Elbow Surg, 11, 315–321.

Sprawls, P. (2000). Magnetic Resonance Imaging [eBook edition]. Medical Physics Pub Corp. http://www.sprawls.org/mripmt/MRI02/index.html

Stanford Health Care. (n.d.). Risks of Magnetic Resonance Imaging (MRI). Retrieved May 08, 2021, from https://stanfordhealthcare.org/medical-tests/m/mri/risk-factors.html#:~:text=Because%20radiation%20is%20not%20used,Intracranial%20aneurysm%20clips

Symms, M., Jager, H. R., & Yousry, T. A. (n.d.). A review of structural magnetic resonance neuroimaging. J Neurol Neurosurg Psychiatry, (75), 1234-1244. doi: 10.1136/jnnp.2003.032714

SyntheticMR. (2017, May 2). Study comparing MAGiC to conventional MR imaging published in American Journal of Neuroradiology (AJNR). https://news.cision.com/syntheticmr-ab/r/study-comparing-magic-to-conventional-mr-imaging-published-in-american-journal-of-neuroradiology--aj,c2255384

Tadros, M. Y., & Keriakos, N. N. (2016). Diffusion MRI versus ultrasound in superficial and deep endometriosis. The Egyptian Journal of Radiology and Nuclear Medicine, 47(4), 1765–1771. https://doi.org/10.1016/j.ejrnm.2016.07.011

Tanenbaum, L. N., Tsiouris, A. J., Johnson, A. N., Naidich, T. P., DeLano, M. C., Melhem, E. R., Quarterman, P., Parameswaran, S. X., Shankaranarayanan, A., Goyen, M., & Field, A. S. (2017, June). Synthetic MRI for Clinical Neuroimaging: Results of the Magnetic Resonance Image Compilation (MAGiC) Prospective, Multicenter, Multireader Trial. American Journal of Neuroradiology, 38(6), 1103-1110. National Center for Biotechnology Information. 10.3174/ajnr.A5227

Taubmann, O., Berger, M., Bögel, M., Xia, Y., Balda, M., & Maier, A. (2018). Computed Tomography. In A. Maier, S. Steidl, V. Christlein, & J. Hornegger (Eds.), Medical Imaging Systems: An Introductory Guide. Springer. http://www.ncbi.nlm.nih.gov/books/NBK546157/

Tazegul, G., Etcioglu, E., Yildiz, F., Yildiz, R., & Tuney, D. (2015). Can MRI related patient anxiety be prevented?. Magnetic resonance imaging, 33(1), 180–183. https://doi.org/10.1016/j.mri.2014.08.024

Termine , J. E., & Macchia , R. J. (2005). Raymond V. Damadian, M.D: The MRI and the Controversy of the 2003 Nobel Prize in Physiology. The Journal of Urology, 241-242. Retrieved from https://www.auajournals.org/doi/pdf/10.1016/S0022-5347%2818%2935047-X

Tillier, P., Leclet, H., Malgouyres, A., Laplanche, T., Madoux, M., Picoult, C., Delvalle, A., & Delforge, P. M. (1997). Le comportement psychologique des patients en IRM: analyse, propositions d'amélioration et apport de l'appareillage à aimant ouvert [Psychological behavior of patients in MRI: analysis, proposals for improvement and contribution of open magnet equipment]. Journal de radiologie, 78(6), 433–437.

Tomassini, V., Onesti, E., Mainero, C., Giugni, E., Paolillo, A., Salvetti, M., & Pozzilli, C. (2005). Sex hormones modulate brain damage in multiple sclerosis: MRI evidence. Journal of Neurology, Neurosurgery & Psychiatry, 76(2), 272–275. http://dx.doi.org/10.1136/jnnp.2003.033324

Tretkoff, E. (2006, July). July, 1977: MRI Uses Fundamental Physics for Clinical Diagnosis. American Physical Society. http://www.aps.org/publications/apsnews/200607/history.cfm

U.S. Food and Drug Administration. (2017, September 12). MRI Benefits and Risks. FDA. https://www.fda.gov/radiation-emitting-products/mri-magnetic-resonance-imaging/benefits-and-risks.

U.S. Geological Survey. (n.d.). The water in you: Water and the human body. https://www.usgs.gov/special-topic/water-science-school/science/water-you-water-and-human-body?qt-science_center_objects=0#

Ullrich, P. (n.d.). Indications and Contraindications for an MRI Scan. SPINE-health. https://www.spine-health.com/treatment/diagnostic-tests/indications-and-contraindications-mri-scan

University of Washington. (2014). Featured History: Magnetic Resonance Imaging. Retrieved from University of Washington - Department of Radiology : https://rad.washington.edu/blog/featured-history-magnetic-resonance-imaging/

Vadera, S., & Bashir, U. (2021). Echo planar imaging. Retrieved from Radiopaedia : https://radiopaedia.org/articles/echo-planar-imaging-1?lang=us

Van Beek, E. J. R., Kuhl, C., Anzai, Y., Desmond, P., Ehman, R. L., Gong, Q., ... Wang, M. (2018). Value of MRI in medicine: More than just another test? Journal of Magnetic Resonance Imaging, 49(7). https://doi.org/10.1002/jmri.26211

Van Beek, E. J., Kuhl, C., Anzai, Y., Desmond, P., Ehman, R. L., Gong, Q., & Wang, M. (2019). Value of MRI in medicine: More than just another test?. Journal of Magnetic Resonance Imaging, 49(7): e14–e25. https://doi.org/10.1002/jmri.26211

Vogel, G. (2003). Image-Conscious Medicine Nobel. (American Association for the Advancement of Science) Retrieved from Science Magazine: https://www.sciencemag.org/news/2003/10/image-conscious-medicine-nobel

Wade, N. (2003). American and Briton Win Nobel for Using Chemists' Test for M.R.I.'s. Retrieved from the New York Times: https://www.nytimes.com/2003/10/07/us/american-and briton-win-nobel-for-using-chemists-test-for-mri-s.html

Wakefield, J. (2000). The "Indomitable" MRI . Retrieved from Smithsonian Magazine : https://www.smithsonianmag.com/science-nature/the-indomitable-mri-29126670/?page=1%3

Wald, L. L. (2019, September). Ultimate MRI. J Magn Reson., (306), 139-144. doi:10.1016/j.jmr.2019.07.016.

Walworth, D. D. (2010). Effect of Live Music Therapy for Patients Undergoing Magnetic Resonance Imaging. Journal of Music Therapy, 47(4), 335–350.

Warner, E., Plewes, D. B., Shumak, R. S., Catzavelos, G. C., Di Prospero, L. S., Yaffe, M. J., ... Narod, S. A. (2001). Comparison of Breast Magnetic Resonance Imaging, Mammography, and Ultrasound for Surveillance of Women at High Risk for Hereditary Breast Cancer. Journal of Clinical Oncology, 19(15), 3524–3531. https://doi.org/10.1200/jco.2001.19.15.3524

Weisstanner, C., Gruber, G. M., Brugger, P. C., Mitter, C., Diogo, M. C., Kasprian, G., & Prayer, D. (2017). Fetal MRI at 3T—ready for routine use?. The British Journal of Radiology, 90(1069), 20160362. https://doi.org/10.1259/bjr.20160362

White, L.M., Kim, J.K., Mehta, M., Merchant, N., Schweitzer, M. E., Morrison, W. B., Hutchison, C. R., & Gross, A. E. (2000). Complications of total hip arthroplasty: MR imaging - initial experience. Radiology, 215(1), 254–262.

Williams, S., & Grimm, H. (2014). Gadolinium Toxicity - The Lighthouse Project. Gadolinium Toxicity. https://gadoliniumtoxicity.com/.

Williamson, I. O., Elison, J. T., Wolff, J. J., & Runge, C. F. (2020). Cost-effectiveness of MRI-based identification of presymptomatic autism in a high-risk population. Frontiers in Psychiatry, 11, 60. https://doi.org/10.3389/fpsyt.2020.00060

World Health Organization. (2016, April 29). Ionizing radiation, health effects and protective measures. https://www.who.int/news-room/fact-sheets/detail/ionizing-radiation-health-effects-and-protective-measures.

World Nuclear Association. (2020, April). Chernobyl Accident. World Nuclear Association. https://www.world-nuclear.org/information-library/safety-and-security/safety-of-plants/chernobyl-accident.aspx.

X-rays. (n.d.). Retrieved May 4, 2021, from https://www.nibib.nih.gov/science-education/science-topics/x-rays.

Xiao, Y., Paudel, R., Liu J., Ma, C., Zhang, Z., & Zhou, S. (2016). MRI contrast agents: Classification and application (Review). International Journal of Molecular Medicine, 38(5), 1319–1326. https://doi.org/10.3892/ijmm.2016.2744

Yim, S. B., Chung, Y., Chung, P. W., & Won, Y. S. (2018, October 24). Effect of Lumbar Drainage on Incidence of Shunt-Dependent Hydrocephalus in Patients with Subarachnoid Hemorrhage Due to Ruptured Aneurysm Who Underwent Coil Embolization. The Nerve, 4(2), 72-76. 10.21129/nerve.2018.4.2.72

Yoon, J. H., Lee, J. M., Suh, K. S., Lee, K. W., Yi, N. J., Lee, K. B., & Choi, B. I. (2015). Combined use of MR fat quantification and MR elastography in living liver donors: can it reduce the need for preoperative liver biopsy?. Radiology, 276(2), 453–464. https://doi.org/10.1148/radiol.15140908

Zagorchev, L., Oses, P., Zhuang, Z. W., Moodie, K., Mulligan-Kehoe, M. J., Simons, M., & Couffinhal, T. (2010). Micro computed tomography for vascular exploration. Journal of Angiogenesis Research, 2, 7. https://doi.org/10.1186/2040-2384-2-7

Zagoudis, J. (2018, March 30). Will Cardiac MRI Expand? Diagnostic and Interventional Cardiology. https://www.dicardiology.com/article/will-cardiac-mri-expand

Zikria, J. F., Machnicki, S., Rhim, E., Bhatti, T., & Graham, R. E. (2011). MRI of patients with cardiac pacemakers: a review of the medical literature. AJR Am J Roentgenol, 196, 390–401.

www.ingramcontent.com/pod-product-compliance
Lightning Source LLC
Chambersburg PA
CBHW021817190326
41518CB00007B/627